Turn Over The Ancient Leaf

How a Schedule 1 "drug" can help save lives

Turn Over The Ancient Leaf
How a Schedule 1 "drug" can help save lives.

Thomas Henry Crinstam

Crinstam Publishing
2015

First Printing: 2015

ISBN 978-1-329-56596-8

Crinstam Publishing
P.O. Box 254
Magnetawan, Ontario P0A 1P0

Dedication

This book is dedicated to the two most special people in my life.

For my wonderful daughter Rowan, for whom I hope the future will be one that is based on what is right, as opposed to what is convenient or profitable. Wherever your path leads you, walk it with your head held high and your heart open, for you always carry my love with you.

To my amazing partner Nicole, for putting up with me through many years of ups and downs, while I work, not just on my writing, but also as I tried to hold the fractured pieces of who I am together. She also is my main proof-reader and co-editor.

Thank you to you both. Without your support and patience, I would have never made it this far.

A special thank you goes out to those that I wrote about here and all those who are standing up for what is right.

Contents

Foreword

Hello, Tom here. This is the second edition of my book. When I first started writing the book, I was in a place in my life where although not good, it wasn't bad. By the time I got around to finishing it though, things had taken a drastic turn for the worse. So, there were parts of the book, especially where I talked about myself, that I was being quite negative.

Since then though, a lot has changed. First and foremost is that we, Canada, had our federal election in which we elected a new Liberal Party government with Mr. Justin Trudeau at it's head. Not only is this good in the general sense that it means we have a bright and hopeful future compared to the last ten years, but also because one of their promises even before the election, was to legalize Cannabis.

Details still need to be worked out of course, so I still hope those he involves in the decision on how to do it recognize that first and foremost, it is a plant. A plant that despite the last century of misinformation isn't just 'not harmful' but is actually beneficial both medically and socially.

The battle to legalize Cannabis will not be over until people can choose to grow it, use it, make it into stuff and share it, just as you can with something like tomatoes, for example. There should be no restrictions on the personal aspects of it. When it comes to commercial enterprises, whether that is growing it to sell parts of the plant, things made from the plant or even plants and seeds, by all means, make regulations and fit it into the tax code. Just as we do with any other product.

Can you imagine being told that you need to ask permission to grow tomatoes or to turn those tomatoes into sauce? How about that you couldn't give a jar to your neighbor? Or have a BBQ and share food made with your tomatoes. Wouldn't that seem insane?

As for the false rumors of it being harmful, there was a time when people thought tomatoes would kill you. Let that sink in for a minute.

There is no argument that can be made about needing security for home gardens of Cannabis that makes any sense in a society that doesn't consider the plant itself or people using it to be a crime; or at least no more than a food type garden. When was the last time you heard of armed gang members threatening someone over their tomatoes? There is no profit in a black market for something that people can grow and use freely. Sure, there will be cases where peoples gardens might get raided for really awesome bud, just the same as someone might raid someone's tomato plants if they see a really amazing tomato. It isn't right, but it isn't to the scope of being criminal or requiring laws in place. If someone were to make an ass of themself, we already have laws to deal with it in place ranging from trespassing to vandalism or theft.

There should also be no discussion of whether or not allowing people to grow and share their own will harm the commercial side of it. We can grow tomatoes, but how many of us actually do? Even for those that do, they end up buying others anyway, even though we could all share them freely if we wanted to.

As for the book, everything I say about this wonderful plant stands as I wrote it, after all, it was written to share this information and get the ancient leaf turned over for everyone, not just here in Canada.

I have, however, removed one chapter about an organization, as they have seemed to become involved in a different cause that has nothing to do with Cannabis legalization.

We have a new day, a better day ahead of us. Not just in Canada though, because our actions here, will help everyone's efforts in their own countries.

The other changes, which I mentioned earlier? I address them at the end of the book.

Turn Over The Ancient Leaf

Introduction

All across the world, people are re-discovering a simple plant that although never actually lost, its significance was turned upside down. It was demonized and used in a racially charged, political game of power and control that then turned into a massive cash cow for governments and corporations.

For thousands upon thousands of years, up until the mid 1930's, it was used extensively in the medical field, both in the prevention and treatment of illnesses, as well as for wounds and other physical, mental and emotional issues.

The demonization and criminalization of this plant has led to the largest incarceration of people ever in the history of the planet, as well as one of the single largest wastes of resources. Instead of using this plant in the thousands of ways it could have been to improve the life of everyone, the exact opposite has happened and everyone has suffered.

Now, despite eighty years of repression, people are rediscovering that not only is it not a 'bad drug', it is simply a plant. However, unlike a lot of other things in our world (both legal and illegal), it has no harmful effects, at least outside of the laws that have been made against it.

People (a small fraction of which you will meet in this book) from across the world are finding out that in some cases it is the only thing that helps save the life of themselves or a loved one. They and the organizations that support them are fighting to change the laws to get this plant legalized in one fashion or another. The wind is blowing, and soon it will turn over this ancient leaf and a new, healthier world will start to emerge.

As with everything I write and say, I encourage you not to accept any of it on faith. Use what you find here as a starting point and catalyst to look for your own information. Do your own research into everything that interests you and come to your own, informed conclusions.

These are mine.

As for who I am, I am 44 years old and I am alive, thanks in a large part to Cannabis. Four years ago, I was diagnosed with stage 3 malignant melanoma after I had a mole removed and biopsied. I was told by my oncologist, who is supposedly one of the top melanoma experts in Canada, that there is no treatment that works on this type of melanoma.

He did suggest trying Interferon, but wouldn't actually recommend it because he said he knew it wouldn't do anything. His actual words were; "I don't recommend this, but I suggest you start on a round of Interferon, just in case it may help". After doing some research and finding out that Interferon is known for turning normally calm and rational people into raving lunatics who destroy their own lives and that of their loved ones, I decided it just wasn't something I wanted to try as I am also bi-polar, which means I already have emotional issues. Taking a pharmaceutical that is known to cause extreme emotional issues not only seems like a bad idea, I really think that it should be cause for a review of that doctor's credentials.

I have done the pharmaceutical route for my bi-polar disorder and it did not go well. They tried many different pills with results that, at best, did nothing and in most cases, caused all sorts of issues that were worse than the bi-polar disorder in the first place. That's saying a lot, because it's not like it's a picnic to start with. Seeing as those pills didn't do what they wanted them to, they then tried combining them in different and interesting ways, thinking that perhaps a combination of them might do something good.

Does it sound like they were experimenting through trial and error to you? That's what it sounded like to me too. To be clear, if there were no other options and the absolute only thing that would help was a pharmaceutical pill, I would take it. However, there is an alternative; one that not only works, but that doesn't have the extremely long list of negative side effects (such as cancer, liver damage and suicide) that are associated with a lot of the pharmaceuticals they offer.

The fact that not only did none of the doctors suggest trying this plant before the costly pharmaceuticals that have seriously risky side effects, but that they specifically warned me against it, started me off on the idea that I should do my own research and make my own decisions.

That research led me to believe that most doctors, whether through conscious effort, inadequate training or just being trapped into a profit geared system, are not doing what is best for patients.

So, one would think that when I was diagnosed with cancer, I would automatically research it on my own right away. However, the doctors and the medical system are very good at terrifying people into submission. They don't just tell you that you have cancer, they tell you that you have limited options, completely ignoring the ones that they (or the corporation they work for), don't make money off of and in most cases, refuse to even discuss. It's just "do as we say, before you die!"

So I did go for surgery to get a six inch chunk of my back removed, along with some lymph nodes, to see if it had spread at all. This was in spite of the fact that the doctors already knew that I had lumps on my adrenal glands that they hadn't told me about at the time; A 3.5 cm tumor on one adrenal gland, and a 2 cm one on the other. After I found out about these other lumps, through an accidental slip by one of the doctors, they said that they were benign and nothing to worry about in terms of being cancerous.

Just as a side note on that part. The adrenal glands are mainly responsible for issues like stress and emotional control. Knowing this made me wonder if that is perhaps something to do with my issues such as bi-polar disorder, anxiety and such that I have experienced most of my life. Despite the fact, that in their efforts to convince me they weren't cancerous, they made it quite clear I likely had these lumps most of my life, the same as my emotional issues, they brushed off any possible significance or connection.

Needless to say, I don't know how to take any of that, then or now. If they were benign, why didn't they tell me about them right away? Why didn't they supply the records of the tests, even after I made an issue of requesting them through the records department and during a personal, face to face meeting with four 'heads of departments', including the CEO of the hospital? I still don't know the answers to those questions and don't really care anymore, as they have completely destroyed any faith I have in the things they say.

Let's go back to my cancer diagnosis. I started off researching and I came across a website talking about another man's experiences with cancer and what he did to fight and beat it. His name is Rick Simpson. You will read more about him later on. The short version of it is that he started making oil out of cannabis buds and it cured his cancer. He shared the information for free with others and it cured their cancer and other ailments. Since then, it has become known as Rick Simpson Oil and he has been sharing what he learned with the world for free and encouraging people to learn how to make the oil for their own ailments.

I tried to talk to my oncologist about Cannabis and cancer... his response was to mumble he didn't know anything about it and get up and walk away. The next time I brought it up, he again repeated he knew nothing about it, but that "there was *nothing* that would help me". He said I would have new tumors pop up and grow rapidly and that the only thing anyone could do, was to cut them out as they appeared and hope for the best. Statistically speaking, if I had maintained treatments and surgery through the medical system, by the end of June 2016, I have a 23% chance of being alive.

So, there I was. On one hand, faced with the medical community, including specialists in malignant melanoma, telling me that statistically speaking, I only had a 23% chance to be alive 5 years after diagnosis, and that there was nothing useful they could do to help me, and on the other hand, a complete stranger on the internet telling me (or more specifically, the whole world) I could save my own life with a plant.

Needless to say, I figured it wouldn't hurt to at least try what the stranger suggested.

So, I did... and I haven't looked back with regret on that decision at all. I got some cannabis, made some oil and started eating it, as I had no visible tumors at the time.

Now, here is the sticky part of it... I don't know for a fact that it cured my cancer. What I do know is that despite the warnings that I would have new tumors appear and spread rapidly all over my body from four years ago on, I haven't had a single one appear. I have however, let fear temporarily convince me otherwise on a couple of occasions, due to the paranoia instilled by the medical system.

Since that time, there have been more and more scientific studies, research papers and anecdotal evidence that show the positive effects of Cannabis on many different things that ail us. Even during the 70's, when the false propaganda machine was in full swing against Cannabis, a study came out that showed that THC caused cancer cells to turn off and die without harming healthy ones in any manner. A radio report at the time on it figures into the story of how Rick got involved with Cannabis and Oil.

These days a quick search on the internet will quickly lead one to all manners of stories and research along these lines. Unfortunately, there is a lot of confusion surrounding it all. There are different corporations and companies that are trying to corner the market on specific patentable aspects of the plant and they are fighting against anything else interfering. You have private scammers trying to make money who are doing the same thing and worse. Then there are the law enforcement agencies that will lose millions (or billions) of dollars when it is legalized, fighting for their jobs, spreading more lies and misinformation. Of course, there are also the doctors and charity organizations that become almost pointless in the face of this simple plant, not only fighting for their jobs, but trying very hard to not admit they have been killing people for profit for decades, that have all contributed to this confusion.

Then, there are the patients themselves that are proving the different healing aspects of the benefits of this plant thinking that it works for what they need, but still assuming all the other lies about it are true. An example of this is people admitting it helps them for one thing, but that they worry about smoking it and getting cancer. The latest studies that are being done, show that not only does smoking Cannabis not increase the likelihood of cancer (including lung), it actually reduces the rates of it. Even the older studies that showed there was a negative connection, have since been looked at and when they actually considered other factors, like tobacco use, the connection to smoking Cannabis disappeared.

So, here in this book, are a variety of different stories from around the world of people using cannabis for different diseases or ailments that we suffer from. Read their stories and check them out; do your own research and hopefully, when you are done you will realize that it's way past time for this plant to be legalized completely. Not just for one disease or disorder, or for certain companies to make money off of, or even just medical purposes, it should be legalized and completely deregulated in every fashion, for any and all personal uses.

For those that seem to miss the point, we already have laws in place that control and regulate the commercial production and sale of pretty much everything else. Those same laws do, or at least can apply the same. Just as if you want to grow Tomatoes commercially you need to follow certain health, safety and commercial regulations, you would have to do the same.

However, just as you can grow tomatoes at home and share them with your friends, make them into sauce and other things without being interfered with by the government, so should you be able to with Cannabis.

It's time to turn over the ancient leaf.

Chapter 1: Laws

Laws should protect a person's basic rights from being taken away or restricted for the benefit of others.

Few of us ever stop to think about laws in anything more than a general abstract manner, at least until we find ourselves on the other side of that wavy legal line. In general, we tend to want to think that laws are meant to protect us from the 'bad guys' and leave it at that. As nice as it would be if that were true, it unfortunately isn't. They are made for a wide variety of, often contradictory reasons. Worse than that, they are sometimes even made based on completely false reasons.

There are many different types of laws; common, civil, aboriginal, parliamentary, legislative and so on as examples, with different countries having their own versions. The relative values of these various systems change according to circumstances surrounding them, usually related to money and power. Adding to the confusion is that a lot of people consider laws to be immutable, even though they are actually undergoing a constant change. You hear people say things like, "Well, that's the law", as if that is all that needs to be said on a subject, without ever considering whether or not that law is just, fair or even reasonable. Sometimes even when they understand a law is wrong, they still just accept that it is there and try to work around it or just simply ignore it.

However, each time a law is enforced, defined or redefined by courts, it has an effect on future cases and changes the way old laws are interpreted, in some cases striking them down completely. New issues are being written up every day, such as ones regarding electronic communications, something that never existed a few decades ago. So there are new laws coming in, sometimes under pressure from the public, other times from private interest groups, all on a constant basis.

Then of course, there is the fact that not only do our societies change over time, but as individuals gain a more public voice, laws can shift and are changed to better reflect the will of the people, rather than the will of those at the top of society.

Over 1600 years ago, Aurelius Augustinus Hipponensis, Saint Augustine of Hippo, wrote words that have been quoted, and at times misquoted, ever since;

"A law which is not just does not seem to me to be a law".

Forms of this quote have been used by many since that time to help support the notion that laws should be just and fair and that when they aren't, you should stand up and fight them.

In 1963, Martin Luther King, Jr. wrote in his letter from the Birmingham Jail;

"I would be the first to advocate obeying just laws. One has not only a legal, but a moral responsibility to obey just laws. Conversely, one has a moral responsibility to disobey unjust laws. I would agree with St. Augustine that "an unjust law is no law at all"."

Throughout history, people have recognized that there are times when unjust laws are created and supported that violate the rights we should all share. Some of the worst atrocities we have ever committed against one another were falsely justified by the idea, codified in law, that all people do not have the same rights. There have been (and still are) laws that reduce a person's value on the basis of random factors of their birth as well as choices they make that affect nobody else.

Obviously this isn't just a recent thing of course; going back to the oldest known code of laws, The Code of Ur-Nammu, there have been laws that support the idea that some people's rights are more important than others.

Thankfully, there have been people who have stood up to power even when it put their own freedom and sometimes their very lives at risk, who ended up getting changes happening to better reflect true justice for all.

Franklin Roosevelt wrote;

"Freedom means the supremacy of human rights everywhere. Our support goes to those who struggle to gain those rights and

keep them. Our strength is our unity of purpose. To that high concept there can be no end save victory."

'To gain those rights' requires that they be recognized in law and that they are applied to everyone equally. When laws are made that allow our basic rights to be violated, everyone should stand to oppose them, because you never know when it will be your turn to have your rights taken away to benefit someone else.

As the poem based on Pastor Martin Niemöller's words says;

First they came for the Communists, and I did not speak out—

Because I was not a Communist.

Then they came for the Trade Unionists, and I did not speak out—

Because I was not a Trade Unionist.

Then they came for the Jews, and I did not speak out—

Because I was not a Jew.

Then they came for me—and there was no one left to speak for me.

The point to this is that we should not accept that laws are right, just because they exist. They cannot be used as an argument against doing the thing they are about without other, openly public reasons. I say they cannot, but I should be saying they should not be, because they are used that way, very often right now.

It is very easy to check the validity of a statement such as 'It must be bad/good because there is a law against/for it'. If you were alive back in the days of slavery, would you say "It's okay to whip that person to death because the law says their skin color is wrong"? Would you have fought against women being recognized as people, just because the law said they weren't? How about laws that allowed feudal lords to have sex with newly married women before their husbands did? Would you have supported these laws, just because they were laws?

Laws can be, and often are, made for reasons that have nothing to do with right and wrong. When this happens, only by having people stand up against them can they be changed.

There are many humor sites on the internet that highlight old laws. Go check them out. You will find many that you will know without a doubt, should never have been laws, yet they not only existed, but in some cases still do and are on occasion tossed about as a 'right' even though they usually fall to any challenge.

Unfortunately, it isn't always funny, in fact, sometimes it is downright horrifying. Laws shouldn't be horrifying, the subject of comedy or be used to harm each other.

They should be protecting each person's rights equally, so that we may all share in life, liberty and justice, which are not a nationalistic ideal, but one that has shone through the leaves of time throughout history.

What are you doing to help ensure that all of our future children grow up in a world that is governed by laws that make sense and protect each other?

Chapter 2: Corporatism

A corporation is structured to limit the liability for its actions from the people who are responsible for making and allowing the decisions. In other words, it allows those who run and invest in a corporation to say "It's not my fault" and have it stand up in court.

Stop and think about that. We raise our children to be responsible for their actions; we tell them that they are the ones who are responsible for the choices they make. Then when they turn into adults, it becomes, 'start a corporation, limit your liabilities'; after all, if you make a bad decision, it shouldn't cost you everything you own and worked hard for, right? Even if you don't have enough to pay the costs of the decisions you made and it costs other people everything they own and perhaps their very lives, you shouldn't have to start over at zero; or at least that is the basis for limiting people's liability through a corporate structure.

At the heart of it, that is one of the main points of a corporation; to allow people to make bad decisions and not face the consequences of those decisions. It of course, does nothing to make sure the benefits are shared equally by those who have to bear the costs of it. This of course, ties into the next important point. That is the fact that, regardless of other factors, because there is no liability, a corporation's number one main reason for being, is to make a profit for those who are invested in it at all costs.

Everything else comes secondary to that.

Ideas such as responsibility and accountability are replaced with deniability and subterfuge. It allows them to make decisions that balance the cost of a person's life (note, that is not the 'value' of) vs. the profit they will make.

Did you have an invisible friend growing up that you blamed everything on? Well, when some people grew up, so do their invisible friends and they start a corporation.

When you incorporate and your invisible friend becomes the owner, you are no longer personally responsible for your mistakes.

Your invisible friend is, so instead of you losing everything you own, they do. Of course, all they own is the corporation. They don't actually have a house or a car or anything else to lose so it doesn't matter what they do.

If there isn't enough to go around, that is just too bad. The corporation's investors and executives could be sitting back in billion dollar mansions, enjoying the lap of luxury while people are dying because of the actions the corporation took and it just simply doesn't matter. The fact that you buying their products or services is what let them afford their luxurious life, is irrelevant because the invisible friend is the 'owner' and in the worst case scenario, the corporation goes into bankruptcy and the investors and executives walk away with a loss limited to the actual amount they had into the corporation.

Of course, there are layers of protection for the corporation itself as well, because let's face it, if they can avoid losing the corporation, it will save them the time and effort in starting a new one. For example, the corporations use their money to lobby, bribe and blackmail to get the laws shifted in their favor. One such way is by placing limits on the amount of damages that they have to pay for a mistake. We see that happen in many cases, but most notably in environmental and health damage cases.

You see, even the invisible friend is protected this way and they can continue on making bad decisions. These corporations will be recording profits in the billions of dollars while the planet dies as a result of their decisions and the rest of us pay the costs of it.

All that matters is that at the end of the day, they show a profit this quarter. Everything and anything they do to run a corporation is secondary to that bottom line. They can kill, maim, and torture people, wipe out whole species, destroy the planet and none of it means anything as long as that bottom line is written in black; the bigger the number, the better.

Of course, those corporations and their main investors are the ones with the most money and power ; so how we fix it, isn't an easy question to answer and the answer is likely going to end up being a lot less pleasant then we could all wish. It's especially bad when you consider that some countries have come so strongly under corporate influence, that corporations are now considered actual people under the law.

Knowing how these things happen though is only part of the first step to changing it. Just suggesting what one of the problems we face is and not actually having any suggestions on possible solutions wouldn't be nearly as helpful as if we could come up with some possible solutions.

The way I see it is that one of the main things that has to be done is to either do away with the idea of limited liability for corporations and their investors, or making the public (through full voting shares held by the government) a 50% owner in every corporation with full public disclosure on everything they do, including research. Now, before a bunch of people throw the book in the fire let me explain the point here.

As things stand, when the liability of any corporation is limited to less than the current, ongoing and future costs of their actions, it is everyone who pays for the remainder of the costs. There is no magical force that makes the costs disappear just because a law says a corporation doesn't have to pay the full costs of their decisions. It just gets spread out to every single person around, whether they are invested in the corporation or have chosen not to invest in any corporate ventures at all or even if they on the brink of starvation without enough money for food and shelter. So, as we automatically have to pick up the tab for corporation's mistakes, we should also have a share in the profits they make off it.

As for the executive branch of a corporation, which includes any major investors (over 5% of the stock), their personal wealth should be linked to the well being of the corporation. The 'golden parachute' should become representative of a clause that allows them to keep the average net worth of the citizens around them (i.e., of the country they are in). That way, when they make decisions, they are aware that they will be personally responsible for their actions including investments, as well as still providing them with a buffer from losing everything due to a corporate liability beyond their control.

Also, as part of the requirements to incorporate, the applicant should have to show a business plan that covers normal operations as well as a long term overview of at least fifty years, or the length of a cycle if something they are using has a lifetime longer than that. As an example of that, if the corporation wanted to do something that involved them cutting down a California Redwood tree with a lifespan of up to two thousand years, they should at least have a concept of how they are going to compensate for their effects over those two thousand years.

To those who would argue that the government shouldn't be involved in their business and shouldn't get half their profits, I say, 'don't incorporate your business'. You take all the risks, act in a responsible manner and pay for the total cost of your mistakes and there is no need for the government to be involved in this manner. If you want to spread the risks of your decisions out to others though, then you should pay the costs of it. Also, the 'government', although it does need to be fixed, is the people's representative. The more revenue they receive from the 'peoples' share of corporations, the less they have to look for, from the people.

Although this wouldn't stop a corporation from making all mistakes or bad decisions, it would have the effect of giving the elected government the power to outright stop them from doing things that are blatantly wrong just because they are profitable and make them realize that they do bear the brunt of the risks for their own decisions.

Chapter 3: Society

For the human society to succeed, we must recognize each other as humans, not because we are all the same, but precisely because we aren't.

A lot of the time, people seem to forget that our society is actually made up of everyone from across the world. It isn't a coherent and cohesive society for sure, but society it is none the less. We tend to focus on a smaller subset of that when we talk about society though.

Depending on the topic at hand, a person may be speaking of their geographical society (North American), their national society (Canadian), or based on heritage (French-Canadian) and so on, but encompassing them all, is the human society. It shouldn't matter if a person is born in a certain location, if they are close to us or far away, the same color, or speak the same language.

We should consider and treat each other the same as we wish to be treated. When someone loses a child, people around them, even complete strangers will reach out with caring and sympathy. Unless it's on the other side of the world or we can assign them to somebody else's society. Then, it's just too easy to blink and look away, mumbling something dismissive about 'those people'. We are all linked through the simple fact that we all share the same planet, have the same needs to survive and ultimately, depend on each other to, if not actively help us, to not harm us or hinder us in our own journey.

We can see the proof of this especially in the negative things that are happening around the world. Violence in one part of the world causes chaos in another part. Political upheavals in one country bleed over into their neighbours. We have corporations that span the world, feeding off and enforcing imbalances precisely because we do not look around at each other as part of our own sub-societies.

Very few people put any thought into the fact that it is through no effort of their own that they were born where they were. There are no forms you fill out before you are born where you choose your race or nationality. You don't get to choose who your parents are, the color of your skin, your gender or even if you fit the conventional classifications of what gender is. You also don't get to choose how you are raised, treated or even educated as a child. So not only is the location you are born in random, the package you come in, the way you fit in that package and how that package is treated during the most formative of your years, is all totally beyond your control, just like everyone else.

So, regardless of color, gender, location of birth or language, we are all human and all part of one global society. We need to start looking around and accepting each other as we are and respecting that, while we are not all identical, we are all on the same trip together, the one we call life.

Chief Joseph of the Nez Perce people from the Pacific Northwest region of the United States said it well;

> *"Treat all men alike. Give them the same laws. Give them all an even chance to live and grow. All men were made by the same Great Spirit Chief. They are all brothers. The earth is the mother of all people, and all people should have equal rights upon it. You might as well expect all rivers to run backward as that any man who was born a free man should be contented penned up and denied liberty to go where he pleases. ...Let me be a free man, free to travel, free to stop, free to work, free to trade where I choose, free to choose my own teachers, free to follow the religion of my fathers, free to talk, think and act for myself — and I will obey every law or submit to the penalty."*

Just to be clear on that, I take his word 'man' to mean mankind, regardless of how it may have been meant, or taken by others. In other words, it isn't singling out the male for special treatment; it has nothing to do with genders. It is about humankind as a whole. We shouldn't focus on matters that are only close to us, either in the physical sense or dear to our heart. For if there is one person left anywhere who is being treated as less than the rest, then the rest are not safe from it happening to them down the line.

We must stop putting blinders on to block out other people's problems because at the heart of it all, their problems are our problems. The idea that there must be an 'Us' or a 'Them' is not only false, but it causes a never ending cycle of conflict, because there will always be someone to call 'Them'.

When we look for things that separate us from being one, we are only isolating ourselves. Each one of us is a unique individual, so we will always find a difference in others when we look.

Instead of thinking of ourselves in terms of being strangers, we should be looking at each other as brothers and sisters, mothers and fathers, aunts and uncles and so on. I don't necessarily mean that as, "We are all one family", but we are all somebody's family (albeit, some are more intentional than others).

Don't you want others to treat your family well?

When you see someone who needs a hand up, you should ask yourself if you would want someone to help if it was your child or another of your loved ones who was in need. On the bright side of things, we have more tools to communicate with each other than we have ever had before. So now we just have to start using these tools to organize as a coherent society where we can all not only feel free and secure in our own lives, but know that all others feel the same.

We do this by looking past the differences among us and reaching out to help when it is needed. Not as part of a plan to spread a religion, or to promote a product, but to help bring basic rights to everyone the world over. We do it by sending a letter or email of support to someone who is struggling, even a simple "You are not alone" can make a world of difference. Beyond that, write one to your politicians who are supposed to be representing your views, starting at the municipal level right up to the national one. It doesn't matter if they have the power to make a direct change to help or not. As your representatives, they need to know how to represent you. If enough people are telling their representatives that they support an issue, it does filter through to those who can help.

Get online and make a blog, share your own stories, read other people's stories, meet the people who share the world with you. Start reading and sharing informative posts on social media sites. They are more than just platforms to play a game or share cute pictures of animals or interesting combinations of letters. Regardless of who runs them or what they do with the information behind the scenes, they also allow us to communicate across our whole global society. Find a way to become an active member of society, even if it isn't in a physical manner.

This is the way we start to build one single, coherent, human society; one that works. A society where we look after each other and consider each other's basic rights to be of paramount importance in comparison to everything else, including profit.

That would be a society that we could all be proud to call our own; and let's face it, if we, the people don't get together and make a single coherent society that works for us all, there are others who will do it anyway, purely for their own benefit.

Chapter 4: Right and Wrong

A most insidious form of fear is that which masquerades as common sense or even wisdom, condemning as foolish, reckless, insignificant or futile the small, daily acts of courage which help to preserve man's self-respect and inherent human dignity. - Aung San Suu Kyi

There have been many attempts throughout human history to enshrine basic human rights in laws. Now, if you actually stop and think about that statement, you should be able to detect a faint odour about it.

Why would I say something like that?

Should there be a need to enshrine 'basic human rights' in law? Do we really need words on a piece of paper and enforcement to make sure we treat each other with the simple respect we want from them in return? One might think that it's obvious we do need to do it; after all, they come about because people make these laws and commit resources to enforce them. Well, at least when it is convenient to do so. That is where the problem comes in.

The UN adopted "The Universal Declaration Of Human Rights", of which this is the preamble;

Whereas recognition of the inherent dignity and of the equal and inalienable rights of all members of the human family is the foundation of freedom, justice and peace in the world,

Whereas disregard and contempt for human rights have resulted in barbarous acts which have outraged the conscience of mankind, and the advent of a world in which human beings shall enjoy freedom of speech and belief and freedom from fear and want has been proclaimed as the highest aspiration of the common people,

Whereas it is essential, if man is not to be compelled to have recourse, as a last resort, to rebellion against tyranny and oppression, that human rights should be protected by the rule of law,

Whereas it is essential to promote the development of friendly relations between nations,

Whereas the peoples of the United Nations have in the Charter reaffirmed their faith in fundamental human rights, in the dignity and worth of the human person and in the equal rights of men and women and have determined to promote social progress and better standards of life in larger freedom,

Whereas Member States have pledged themselves to achieve, in co-operation with the United Nations, the promotion of universal respect for and observance of human rights and fundamental freedoms,

Whereas a common understanding of these rights and freedoms is of the greatest importance for the full realization of this pledge,

Now, Therefore THE GENERAL ASSEMBLY proclaims

THIS UNIVERSAL DECLARATION OF HUMAN RIGHTS

as a common standard of achievement for all peoples and all nations, to the end that every individual and every organ of society, keeping this Declaration constantly in mind, shall strive by teaching and education to promote respect for these rights and freedoms and by progressive measures, national and international, to secure their universal and effective recognition and observance, both among the peoples of Member States themselves and among the peoples of territories under their jurisdiction.

In thirty articles, this document lays out how we should treat each other, it isn't perfect by any means, but it could be a start of something great. Unfortunately, it is neither universally accepted nor recognized, nor is it adhered to by those who did recognize and accept it.

Think about how insane that is, what would happen if it didn't exist? Would it change anything? Would people suddenly need more food or air? Would they get sick less often? Would they need more shelter from the elements?

I think if it didn't exist, we would all still be people sharing the same planet with the same needs. Don't get me wrong, I am not saying it is worthless. Having the 'powers that be' pay lip service to basic human rights is better than them ignoring them completely; it's just the idea that we need to make laws or rules to treat each other fairly, right from the start, is a sad reflection on us all.

However, the fact that there is at least a framework for human rights and some form of adherence to it is a start. In fact, it really is the basis for any start that will end up being good for all people.

Forty percent of the world's 'wealth' is held by one percent of the population. If you really think they have done that by luck and chance up until now, you may want to take a look through history. That sixty percent that the ninety nine percent split among themselves, is better than it has even been in a long time. Now, before people think I've lost my mind, I will point out that the last twenty or thirty years in some countries would seem to point to that disparity growing, but in actuality, it is returning to where it was and has been for most of history; at least since we started allowing each other to take away our rights, with a few controlling the majority of currency, power and people.

Although not accurate, the words "let them eat cake" echo through time as a warning to the elite and powerful that there is a limit to how much they can steal from everyone else before they rise up. We, the people have been trying to take back our individual rights going back to the beginning of human societies. We know this because we can see the same type of organizational structures we use, in other primate species where the strong rule over the rest. But by learning to communicate with each other in meaningful manners, we have slowly been gaining the ability to reason and understand that regardless of our baser instincts, we can actually respect each other as free thinking individuals.

The problem is, as we have been struggling to fight our baser instincts, those who were on top by nature of their predatory and aggressive instincts have also been learning and gaining more ability to reason. They don't want to give up the power they have been stealing, for any reason. In fact, their instincts tell them that letting that happen is wrong, because beyond wanting to have enough for themselves; they haven't learned to suppress their baser instincts enough to understand the pain and suffering they are causing.

Now, are those people who are in charge, communicating and cooperating with each other? Of course they are; no question about it.

Are they trying to form a single worldwide society that one of them will control? I don't think so, although I do think there are many individuals among that group that would love to do just that and subsequently declare themselves the 'emperor of the world'. Fortunately for the rest of us, I think you will find that the competition among them to make sure one of the others doesn't become that person; makes it unlikely that they are working cooperatively towards that end. Besides that, with separate nations and sub-societies, they and their small group of 'friends' grow rich off the imbalances and profit off suppressing one group to pay off another group to help do it, or at least look away while it is done.

If it were to become a single world order, then all of them would be subject to the rule of one, and they don't want that. They each want to be at the top of their own pile. Not a single one of them care about the basic rights of others, despite anything they may say. Proof of this fact is that they live in opulent luxury while billions starve. There is no justification that can make it okay for such an imbalance to exist if everyone cared about each other's basic rights. Whether they walk past a starving person or fly overhead in a jet is irrelevant, they are actively ignoring other human being's suffering and dying, in order to enjoy their day and feel important.

By no means am I arguing against capitalism and the concept of some people earning more than others. Some of the concepts of it are actually useful to a society. However, the basic needs of people cannot be a commodity in any type of capitalistic system without denying people their basic rights.

If everyone is enjoying the basic rights that we should all recognize, if everyone has enough food, water and clean air, shelter from the elements, education, healthcare and freedom to communicate with each other and security from direct harm, then there would be nothing wrong with some having more than others, even if they had substantially more. Some people do want more than others, and sometimes they are willing to work harder than others to get it. That is great, it allows for each person to be the individual they are.

It's when we start withholding basic needs from people and using that as a threat in order to control them, that the quest for wealth becomes evil. When one's goal of getting ahead, includes allowances to step on others, take what they have or allow them to die, then these are problems that need to be addressed.

It's when we start claiming that doing those things are right, because those in power passed laws that said they were, that everything in society breaks down. It becomes a weird mix of money, power, personalities, ideologies and personal beliefs that all get tossed into the discussion that add so many nuances to right and wrong, that people stop thinking about it and let others tell them what is best.

To me, right is what you want for your loved ones, wrong is what you wouldn't want for them. Pick any issue you wish and ask yourself, if it was about you or someone you love, how would you feel about it? When I say you or your loved one, I don't mean make it an issue between the two of you in your mind. I mean think of it as if you were the person it was about and separately as if the person it was about, was someone you loved. We will often sacrifice our own happiness and well-being in order to help others we care about or for a short term gain that we feel is worthwhile. We may also allow ourselves to think that we should protect our loved one, even if it's from what they want and feel makes them happy; where we would want to be allowed to make that choice differently for ourselves. When you combine both, it helps get rid of those glitches in the thought process. If you wouldn't want someone to do it to you or to someone you loved, then chances are really good that it is wrong to do, or to allow it to happen to others.

There are some generalities that we should all be able to agree on, once we strip away the distractions. Would you like someone to kill you or your loved one? We will assume the answer to that is no. If it isn't, then the book you require is on a whole different topic. So, in what case is it okay for you to support, in any way, the killing of someone else, unless it is in direct response to a threat to do it to another? Ignoring the issues of defense against direct action, we should consider all killing wrong, regardless of whether that is next door to you, or on the other side of the world, shouldn't we? The single most basic right we all have is the right to life. Unless we are threatening other people's right to life, ours should not be under threat and the same goes for everyone else.

That of course, is one of the more extreme examples. There are others that follow along the same thought process that shouldn't need to be explained out in detail. You wouldn't want someone to rape you or someone you loved, so it is wrong to rape. You wouldn't want to be beaten or harmed as a child or to have a child you love, beaten or harmed, so it is wrong to beat or harm children.

There are also the less extreme examples. Would you like to be told whom you have to have sex with and marry? Would you like to be bought and sold as chattel against your will? Would you like to be locked up for trying to heal people? How about being told you can't try to save your life in a manner that affects nobody else? If you wouldn't want it to happen to you or someone you care about, you shouldn't want it to happen to anyone.

Going further than that, would you want someone to step in to stop one of these types of things from happening to a loved one if you weren't there? Yes? Doesn't that mean you should step in to stop it from happening to others, even if they are someone you don't know; even if it is over an issue that doesn't directly affect you or your loved ones?

Right and wrong are not concepts that should be coming down from above, either spiritually or legally. They should be coming from our hearts. Even if someone isn't a loved one of yours, they are a loved one of somebody and if you would like a world where you can trust a stranger to help someone you love, then you had better help a stranger.

That is what is right.

Chapter 5: We allow

We are always looking for others to blame for the position in which we find ourselves in the world. The sad truth is that everything that is wrong with the world is collectively our own fault.

We allow these things to go on. We are the ones who don't often enough, stand up and say, "Enough is enough". We are the ones who sit back and watch all these horrible things happening and do little or nothing to stop them.

You'll often hear people say things like, "Well, compared to this other person (or company or government, etc) I'm great". The problem with that theory is that it assumes that everyone else is trying as hard, or harder than you are, at being 'good'. It's like the idea that if you surround yourself with people who are fatter then you are, you look skinnier. Then of course, you have to ask yourself, are you going to look for people fatter then you, or are you going to help those close to you become fatter? In a lot of cases, people actually will help those around themselves gain weight, or sabotage their efforts to lose it. After all, if the person you are comparing yourself to, ends up losing more weight than you do and looks skinnier, that would make you look fatter, which was what you wanted to avoid all along. It's so much easier to buy them a box of donuts than it is to find a way to bring your own body back into balance and lose weight. Although it is true that you will look skinnier, you won't actually be skinnier.

Which is better? Hiding an issue by making sure those around you have worse issues, or fixing the issue in yourself? Of course this isn't about being fat, that's just a placeholder for whatever it is that you think is wrong in the world.

How about we start comparing ourselves to what we could be, if we would just try a little harder today, than we did yesterday? I mean this on every level: instead of looking at the person next door and thinking that we are better than they are, look at who we could be, if we just tried a little harder today than we did yesterday. Instead of our towns, cities, provinces (in fact our whole country) looking around and thinking that they are one of the best available, they should look at what they could be if we put some more effort into it. The same applies to everyone and everything equally; we have to stop just, 'getting by'.

I don't know about any of you, but I am really tired of turning on the news and seeing us killing each other in more creative and insidious ways every day. I don't want to hear that we don't care. I don't want to hear that you are better than the guy next door. I want to hear you are a better person today, than you were yesterday. I want a peaceful, 'just' world, and it is possible; we just have to stop allowing ourselves to accept anything less.

There are better alternatives than what we have been doing. Look for them, find them, and don't accept anything less!

If everything is possible, why not choose only the good?

Chapter 6: Natural Health

One of the reasons I am so fond of the natural world, is because it provides the things we need, generally in the form we need them. This isn't a magical or a spiritual statement, nor is it just a 'wild theory' (pun intended).

Regardless of how you believe humans came to be here; whether that is created by a god, planted by aliens or simply nature taking its course, we can pretty much all agree that we have been here for at least 6000 years if you accept some religious teachings or the much higher scientific answer of us (Homo sapiens), evolving over the last 200,000 years.

So, for thousands, if not hundreds of thousands of years, we have been born, lived and died here on this planet (or at least close to it). Up until the last 100 years or so of that time, our survival depended 100% percent on the natural world and what it provided for us.

Our food was organic (there was nothing else), our water and air was clean and we didn't have a constant bombardment of unnatural energy assaulting us constantly and when we got sick or hurt, we used what was surrounding us all, nature.

Now, I don't want people to get the idea that I think there is a natural solution or remedy to everything, especially these days when we have so many unnatural things happening to us. Having said that; I do think that the large majority of things that happen to us, can best be dealt with using something natural, almost as we find it in nature.

I realize that there are some people who may be tempted to just brush this idea off when they see it, but I hope you follow along enough to at least understand how I reached this conclusion, even if you don't reach the same one.

The majority of products that are being produced for the medical industry are simply them isolating a specific compound found in a plant known to help and making a marketable product. Even one of the most common pills you can think of, Aspirin is simply a synthetic form of one of the compounds found in Willow bark. That 'one of' part, combined with the fact that it is concentrated in form, make Aspirin much more potent than Willow bark tea would be and in some cases that is good, but for most purposes people use it for, it is like using a sledge hammer to kill a mosquito. Much like in that scenario though, there is bound to be damage done to other things besides just the mosquito, especially if it landed on your head!

Some will like to point out about this time that Willow bark tea can be harmful. While true, any harm from the acetylsalicylic acid in Willow bark tea, is concentrated in Aspirin form along with the benefits.

Now here's the important thing to keep in mind as it applies to all products sold through our medical systems. It isn't about doing what is best for the people who need medical help, it is about making profit. How much profit can you make from something as cheap as Aspirin? Bayer, the company that makes Aspirin, makes over a billion dollars worth of sales per year. That isn't including the generic versions that are being sold as well.

Just because someone is making a profit though isn't proof that they care about people's health less than they do about profit, but when you combine it with the number of times pharmaceutical corporations have been caught lying, suppressing negative tests and promoting unfounded benefits of their products, it becomes pretty obvious.

What about the thousands of years before pharmaceutical corporations came into existence and started their unrelenting quest for profit? We had shamans, druids, medicine men and women, witches, wise women, hermits and all sorts of other names for natural healers that knew what plants to use for what ailments. They were nothing like what a doctor is though. Whereas a doctor waits for someone to come to them with a problem, these natural healers were a significant part of the social fabric that made up our societies of the time.

They didn't just heal the sick; they helped the healthy stay that way. In spring time they would make up various teas and dishes for the whole community to help wash away the lethargy of winter and prepare the body for more strenuous times. When they saw someone who seemed to be under the weather in some form, they would suggest certain things to eat or not eat or things they could do to help with whatever it was.

Now, obviously there are some things that we have benefitted greatly from in the advance of modern medical knowledge, but this is mostly in the form of being able to perform surgeries and various tests we have now. However, even in those cases, the way they are being pushed is wrong. The whole medical system is set up to suck as much money out of sick people as possible to pay for the years of training and brainwashing that goes into creating a doctor.

Think about this concept for a minute. If you came across a person dying of thirst and you had extra water, would you demand they pay you in some fashion before you helped them? I'll rephrase that to make it an easier question to answer. Would you let them die if they didn't have something that you wanted from them? I'll rephrase it a third way; would you want someone to let your child die, for want of a drink of water because you couldn't pay for it?

The very fact that that very thing is what every member of the medical profession does today, should be enough to make any reasonable person weep. Some will claim that the money is needed to pay for all these advances and so on. Let's think about that shall we? Outside of entertainment and sports, doctors are some of the highest paid people in our society and the higher they are on the pay scale, the fewer hours they work. Who doesn't know of at least one doctor's appointment that was rescheduled because the doctor had a surprise opportunity to go golfing? Do you know who one of the biggest groups of private lenders are in the real estate business? Of course, the answer is doctors. These people are literally getting rich off of sick people's desperation to be healthy. In some cases, they quite openly turn their back on those in need and let them die because the person can't afford to pay them what they want.

Why would so many people who start off wanting to help and heal the sick end up doing things like this? Look at the way they are trained. If any other industry used the long hours of study, work and constant bombardment of information to train their employees, they would be called a brain washing cult. That is not an exaggeration; do your own research into the path to becoming a doctor and compare it to any 'successful' cult type organization and see what you find. I did and I don't understand for a second how it is we have allowed this to happen to an field that is supposed to be helping people.

I don't suggest that I know all the answers or even any for that matter, but I think I may have an idea of what we could do anyway.

If you assume that the system itself isn't going to change very quickly and doctors will continue to be so costly than we need another solution outside of it. How about instead of hiring a new doctor at a hundred thousand dollars a year, we hire a couple of herbalists. Instead of people going to see a doctor about every little scratch, sliver and cough, they first go see the herbalist who will suggest natural remedies to try if they think its appropriate or in the case of something that seems more serious recommends seeing a doctor about it? Allow them to order tests and referrals as well. Not only would it mean that more people could get healthy in a more natural and cheaper manner, it would mean that the doctors would have more time to deal with the more serious cases.

At the same time, we have to get rid of the restrictions about using the natural plants that we have evolved alongside for thousands of years. The only way that nature should be restricted in terms of us using it is if it is done for commercial purposes. Look at Bitter Almonds as an example; in their raw form, they are full of cyanide and must be treated with heat before sale. There is no law that says you can't grow them, have them or even use them yourself (although you shouldn't unless you heat treat them of course); there is nobody sitting in jail because they had a backyard full of trees growing them. Yet they can outright kill you if you eat them raw. Commercially speaking though, there are very specific and rigid laws that say they have to be heat treated before general sale. Even with that though, you can still buy them as speciality items without permits or permission.

So, why do we make laws against people using nature as we have used it for thousands upon thousands of years? How does this make sense, especially concerning parts of nature that we know are helpful?

Chapter 7: Personal Stories

"Storytelling in general is a communal act. Throughout human history, people would gather around, whether by the fire or at a tavern, and tell stories. One person would chime in, then another, maybe someone would repeat a story they heard already but with a different spin. It's a collective process." ~ Joseph Gordon-Levitt

There is no way I could write about all the people around the world that fall into the category of those who are fighting for the basic right to determine their own health care, there are just too many to even consider it. Just trying to list them with no story would take up more room than could ever fit in one book. Even when you focus it down past people who are fighting to choose this specific plant, the field of choices is staggering and it is growing by leaps and bounds every day.

I wanted to have a dozen different stories of people that are fighting for the right to use this one single plant as it helps a wide variety of ailments we suffer from, all around the world. Not only do they offer up direct examples of the issues I am discussing, they also offer you the opportunity to learn about the different aspects of the healing properties of the plant. Unfortunately, as I mentioned earlier, that just wasn't possible for this version. There is a chance of a new version, with your help. I'll explain that later.

Some of the stories are of people who are public about their use, one is not. In the next version, I will include two separate sections, one for those fighting openly and those fighting quietly due to the risks involved. There is no judgment of one being better than another, people have to put their (and their loved one's) health and life above all else. If they are in the position to openly fight, that's great, but even if they aren't, they are still helping, not just themselves, but others as well.

I encourage you to look up the people I highlight here where you can and go find out more for yourself. Some of them are not only fighting an illness at the same time as they are fighting against the unjust laws, but are also struggling to make ends meet. Stand up and support them in any manner you can, because not only do they have right on their side and not only could it help save a life, but you should do it, because you would want someone to help one of your loved ones if they needed it.

I should point out that I have never met any of these people, although I would like to one day. I found out about their stories online and made contact with them for the purposes of this book. In some cases we emailed or sent messages back and forth and sometimes when I was able and couldn't avoid it, we talked on the phone. It's important that this is pointed out because we each have our own abilities and strengths, different manners in which we can help, regardless of our circumstances.

For me, getting out in public and talking to people, attending rallies and participating in things like that is pretty much impossible. To be around more than a couple of people I don't know at the same time, is excruciatingly painful. That doesn't stop me from being able to gather information and points of views, talk to others and present them in a manner that could encourage others who do have the ability to do more to get involved.

I picked a wide range of people, scattered around the world, living in different cultures and struggling with different circumstances. Well, at least I tried to; keep in mind that some countries have more freedom for their people to stand out and as such, gave more of a selection, so there does tend to be a lean towards a smaller selection of locations than I wanted. I tried to do this because it isn't about one country or one culture, about poor people or just the young or old. It is about all of us; each and every one of us.

We are the ones who can do something to help these people as well as ourselves; to help change the systems and the laws to reflect that, when it comes down to it, we should be able to choose what we put in our own bodies.

So, read the stories here, close your eyes and imagine if you were in these people's position. Would you want someone else to speak out on your behalf? Would you like someone to help you? Would you like to know you weren't alone? I know I sure would.

Then, ask yourself the most important question of all:

"What am I going to do to help these people?"

Chapter 8: Tara O'Connell - Dravet Syndrome

"Take her home, love her and maybe if you are lucky she'll get to nine"

That is something that no parent should ever have to hear. Unfortunately, it does happen. It happened to Cheri O'Connell concerning her daughter Tara, when she was seven years old. It wasn't completely unexpected though. Tara's parents, Cheri and David had watched her suffer from the time she had her first seizure at six weeks old, although the diagnosis of Dravet syndrome took years.

For seven years, they tried seventeen different pharmaceutical drug combinations prescribed by doctors. Not only did none of them work, many caused major side effects including a heart murmur, inability to regulate body temperature, speech, cognitive and mobility issues as well as others that likely won't present themselves for years.

She was dying, could barely walk or talk, was unable to toilet herself and slept a similar cycle to a newborn. The prescriptions she was on were leaving her heavily sedated and unable to function or grow, yet she was still having from sixty up to hundreds of seizures a day. That's one every twenty four minutes on a good day, to one every six minutes on a bad one.

Now, as if this wasn't enough turmoil for a family, one of their other children, Sean was diagnosed in 2011 with GEFS+ (Generalized Epilepsy with Febrile Seizures Plus). He was taking epilepsy medications plus sleeping pills and anxiety medication for two years. He couldn't concentrate at school, his hand would shake so bad he wouldn't want to write and things were going downhill.

What would you do? Hopefully, you would do what Cheri did. Read, research and talk to people and look for something that could help her children. What she found changed her and her family's lives forever. On YouTube, she came across a video about a parent in the US using Cannabis to treat his child with Dravet Syndrome. In his case, he went from hundreds of seizures a day to none and was weaned off the prescription drugs that weren't helping anyway.

So, in January of 2013, she started giving Tara liquid marijuana, an extract of the essential oils of the Cannabis plant. Ten days later,

Tara stopped having seizures and had no more for 3 months. She had a small one in April of 2013, and since then, not one single seizure more. Imagine, going from at least sixty seizures a day, to 1 seizure in a year. In March, they started Sean on the Cannabis oil as well. He hasn't had a seizure in 9 months now.

Now, this doesn't mean that their lives have become normal by any means. Tara still has a lot of damage that was done to her from both the seizures and the side effects of the medications that she was prescribed for pretty much her whole life. We're talking about brain and nerve damage from toxic chemicals as well as the damage throughout the rest of her system.

Fortunately, the Cannabis oil isn't just stopping the seizures, although even if that was all it did, it should still be considered a miracle drug, the miracle doesn't end there. It is also helping her body repair itself. Since starting on Cannabis Oil, Tara's overall cognitive performances were significantly improved over the last 12 months according to a neuropsychological assessment report and along with the cessation of seizures; there is a convincing clinical history of improvements in all facets of her presentation. She not only doesn't need her wheelchair anymore, she gets into trouble for riding her bike places she shouldn't be going. Sean, who could barely write (and didn't want to even try anymore), is now not only happily writing, but his other issues have disappeared along with the pharmaceuticals.

Unfortunately, this still doesn't mean all is now right. You see, they are all criminals. Cannabis, in all its forms is illegal there. Despite the fact that the O'Connells have incontrovertible evidence that it, and only it, works, as well as support from their doctors, the government there has refused to even consider changing the laws. The O'Connells have the support of their community as well, including the school that their kids go to and the local healthcare representatives. They haven't been personally threatened with the law, but it is hanging over their heads none the less.

As you might imagine, considering that they are doing this for the sake of their children's very lives, it wasn't a hard choice to save them in spite of any legal risks. It would also be very easy to understand why they would hide what they are doing and try not to attract any attention to themselves.

Not the O'Connells though, they know there are many kids out there that are suffering like theirs were, who could benefit from using Cannabis and they aren't willing to sacrifice them in order to keep quiet for their own sake.

Enter Tara's Law. This is their fight to legalize Cannabis in Australia, for the sake of their children and everyone else's. Instead of hiding and keeping quiet, they have become some of the main faces of the battle to legalize cannabis in Australia, so that their choices and experiences can help other parents in the same position.

After seven years of what can only be described as daily torture for all involved, they were able to finally get to know the person that is their daughter. To see her laugh and play, have fun, and yes, even get upset with her for 'running away' to her friend's house to play. To go from spending every moment (both waking and sleeping) with the fear of losing her in the immediate present and having almost no hope for her future, to seeing a bright and blossoming future for her or at least her being able to experience the joy of being alive isn't something they want to keep for themselves.

They want all parents everywhere to have that same experience. To do that, they are willing to risk speaking out, sharing their story, harassing their politicians and educating everyone about what they have learned. In spite of the fact that to do so, is to admit and shout out that they are breaking a law that could land them in jail.

That makes them all heroes.

Speaking of heroes, there is someone that I haven't mentioned in this as yet. As I was learning about their story and talking to Cheri about it all, my mind kept wondering about their other child.

Jasmine is her name and out of them all, I wanted to give her the chance to speak out in her own words. More then just the typical big sister, she has lost a lot of her own childhood in order to help her family. When other girls were growing up and spending time with their friends, dressing up, playing house and such, she was helping to care for her siblings and her parents. It has had rewards, beyond the joy of helping a loved one, she has won an award as a 'Young Carer" and turned some of her experiences into stories that were published in a book about Young Carers just like herself. I don't think that any of that makes up for her lost childhood moments, but I do think that the compassion, strength of character and heart that this young lady has grown and shown through her experiences should be considered an example for everyone else, from child to adult.

So here, in her own words of what a part of her experience has been.

Hello my name is Jasmine. I have a sister named Tara and she has Dravet Syndrome. She is 8 turning 9. I am a young carer for her and my twin brother Sean. We are turning 12 this year. I have cared for them a lot.

Because I am a carer for my siblings I had a dream to go up in a hot air balloon with my dad. So I signed up to go up in a hot air balloon with "little Dreamers" who grant wishes to kids like me. I got picked out of about 60 people. I was with my young carers group when I found the sign up form and I'm like I will put down my phone number. So when I got home that night I told mum and she said that I could try.

Then about a month later mum got an email saying I got picked to go up in the hot air balloon for FREE. It was awesome but cold. We stayed at a hotel in Melbourne. We just wanted a normal double bedroom but they had run out so they upgraded our room for the same price as the room we wanted. We got at the hotel at midnight and then had to get up at four in the morning.

I will always love my sister and I sometimes give mum a break because Tara sleeps in mums room. I sometimes make her sleep in my room or I sleep in her room.

We used to have a cat named "Yoda", Tara loved her to bits but we also have a dog. The dog now loves it because the cat is not around anymore so Tara gives her more attention

Tara has Dravet Syndrome, but you can't even tell anymore because she is on medical cannabis. She used to have up to 80 seizures every day. Sometimes I got scared when I was little, so I saw a lady who helped me let out my feelings.

My brother Sean has epilepsy and autism. I hate his autism because it makes him smart at maths and I am terrible at maths.

Tara likes lots of pink and Dora. So almost everything in her room is pink and got something to do with Dora.

My favorite things are animals. My favorite colors are pink and purple.

A few years ago the doctor said that Tara might only get to her 7th year and then she might die. But we proved them wrong and she is still with us all because of the medical cannabis.

We all love her to bits. When I was eight mum said I will never get a bunk bed but now Tara doesn't have seizures I might get a bunk bed. Tara has a hospital bed, she loves it cause it goes up and down but the only thing is she doesn't sleep in it!

As with most things, if you want to see the true heart of human strength, compassion and giving, look to the young ones, before they get taught to stop caring for others.

Chapter 9: Zach Lockwood - PTSD

Post-traumatic stress disorder (PTSD) is an anxiety disorder that can develop after a person experiences traumatic events.

Honestly, until I started writing this, that was just about the sum total of my knowledge of it. I knew a few other things, like it affects soldiers and victims of crimes and trauma. I also knew that for a long time it was considered an imaginary thing much like many other illnesses. Beyond that though, I didn't really have an occasion to do a lot of checking into it.

Obviously, this book had changed that and I needed to get some information, so I started reading and talking to people. Regardless of the causes of this disorder or even if the personal causes are unknown, make no mistake, it is a serious and sometimes life long battle that effects millions and millions of people worldwide.

Many people lose this battle, either through direct actions such as suicides, or indirect by taking extreme risks or ignoring warning signs of things. It doesn't only affect the people who actually have it either; it affects everyone around them, most especially those who care about them.

Now, although I won't claim to have some sort of inside secret; due to my own experiences, I think I can describe it in such a way that most people can at least grasp what it is like. Although I never really talk about it much, I am pretty sure this is actually one of my own issues in life. I don't generally say that though, because for the most part, I don't want people who have been through things like combat (in whatever role) to think I am saying I suffer the same things they do. Our nightmares and their effects take our own experiences and use them against us. The experience of actual combat type situations is beyond the scope of ordinary human experience as far as I am concerned. Having said that, I know that for someone to go through worse things than my mind puts itself through and survives; they deserve every respect in the world for it.

Imagine your worst fears, thoughts and imaginations, all rolled into a nightmare you can't wake up from. Now, consider what it would be like to vividly remember that at random times throughout the day. It wouldn't be enjoyable would it? Now, imagine if you couldn't vividly remember it, but felt the same fear and horror anyway over and over again.

Although not meant to be an encompassing view of what it is like to suffer from PTSD, it does give a good start for understanding the effects of it on a person's mind. What would you do if you had that happening to you? What most people do is talk to a doctor and start getting various pharmaceuticals tossed at them, at least if they are believed anyway. On the rare occasion, those pharmaceuticals help, more often than not though, it begins a long process of trial and error with the person already suffering, being used as the guinea pig.

Now, some will say it doesn't matter if you have to go trough dozens of medications if that is what it takes to solve the problem. Unfortunately, that is assuming two things are true that are actually false. The first one being that there is a magic pill that will make the problem disappear and it is just a matter of finding it. Many people go through the full range of pharmaceuticals offered and still find little or no relief. The second one is that these pharmaceuticals aren't causing harm in your body. Look at the list of possible side effects for the pills they offer and keep in mind that it has been shown over and over that all of the major pharmaceutical corporations have lied and hidden negative findings about their products.

So, not only is there a possibility that none of the commercial pharmaceuticals will help a person with PTSD, but they could actually be further harming them.

Now, instead of leaving it to your imagination as to what all this would be like, I'd like to introduce you to Zach Lockwood. In 2005, Zach joined the military and entered the Military Intelligence sector, serving in both Afghanistan and Iraq in a mobile intelligence unit. While unable to pinpoint a specific incident of what it was that caused his PTSD, it shouldn't be to hard to grasp that being in combat with not just your own life, but the lives of friends that you have lived and trained with being at great risk is going to cause more than a little stress in a person's mind. So, after serving for almost 5 years, including a year over his original contract (known as being stop-lossed), he retired back into civilian life, only to discover that his experiences had left him unable to cope with certain situations due to the PTSD.

Sleepless nights, alcohol, cocaine, anything that would blank his mind and allow him to not think for even a little while, took over his life. Seemingly innocent circumstances would snap his mind back into the same frame it was during combat, fight, flight or die. Anything that would make that fade seemed like something to try. Unfortunately, those coping mechanisms were causing more problems than they solved. It took two years before he got to the point that he knew he needed a way to deal with this that wasn't going to cause more, and worse, problems down the line. Fortunately, he wasn't alone in his struggle. Some of the friends he served and fought alongside of were also having the same issue. Some of them tried doing the conventional treatments and ended up, in overwhelming numbers, addicted to the very medications that were supposed to help them; sadly, without any of the benefits they were supposed to bring. Some of them crashed and never recovered. So he started doing research on his own, trying to find a safe and effective way to deal with his problems.

Now, you will find that a lot of professionals in the health industry are stuck on this idea that the best thing they can do for someone is to try to cure their problems. If this was possible in all cases, then it would be a great thing. Unfortunately, they have let their training and egos get the better of them and have forgotten that the patients they are dealing with are real people, with real lives that are being affected by their attempts. To someone suffering, sometimes it is not only more important to find even temporary relief, but it can save their actual lives.

For Zach, as well as many of his fellow PTSD sufferers, this temporary relief comes in the form of Cannabis. While not immediately solving the problems, it does allow the sufferer enough peace of mind that they see hope for the future. Going beyond that, more and more studies are showing that Cannabis may also be able to effectively reverse the effects of PTSD over time.

Unfortunately though, this means that the very thing that they find the most relief from is also causing extra stress, as in a lot of cases, as you have to break the law in order to get and use it.

This type of attitude is the epitome of the saying, "kick them when they are down". We need to stop doing this to each other.

Chapter 10: Robert Platshorn - Former Drug POW

The right to work, I had assumed, was the most precious liberty that man possesses. Man has indeed as much right to work as he has to live, to be free, to own property. The American ideal was stated by Emerson in his essay on Politics, 'A man has a right to be employed, to be trusted, to be loved, to be reversed.' It does many men little good to stay alive and free and propertied, if they cannot work. To work means to eat. It also means to live. For many it would be better to work in jail, than to sit idle on the curb. The great values of freedom are in the opportunities afforded man to press to new horizons, to pit his strength against the forces of nature, to match skills with his fellow man. -William O. Douglas, Associate Justice of the Supreme Court of the United States

United States – Robert Platshorn, actor, pitchman, successful entrepreneur in America and England, Bull Fighter in Spain, TV producer, pilot, big game fisherman, and America's longest serving non-violent marijuana prisoner.

In the 70's, he and his crew, labeled as the Black Tuna Gang by the media and law enforcement, smuggled Cannabis into Florida.

After retiring and getting out of the business, Robert was indicted in 1979. The US Attorney General, Griffin Bell, held a press conference in Washington, DC where he called The Black Tuna Gang, the "slickest, most sophisticated pot smugglers of the 70's."

Following a controversial trial where the DEA and FBI tried (and failed) to entrap members by getting them to bribe jurors and falsely accused them of planning to assassinate the judge, he spent the next 29 years in federal prison, becoming the longest serving nonviolent marijuana offender in the United States.

He was released in 2008 and now lives in Florida with his wife Lynne, working on his writing, videos and speaking tours regarding the medical benefits of Cannabis.

Now, I'm sure there are some people who are confused at this point, as it seems like a simple case of someone who broke the law, served his sentence and moved on. I'm sure there would be nobody happier in the world than him if that were true.

Even if you don't consider the war on drugs to be the biggest travesty ever embarked upon, I'm sure you will agree with one thing: serving more time in jail for selling a plant, without violence being involved, than most murderers and child molesters do, is wrong. That never should have happened.

Considering it was over a plant that should have never been outlawed in the first place and is on the verge of being legalized across the world, what he did shouldn't have been a crime, any more than importing coffee or bananas are crimes.

However, that is the past and we can't change it. What is important now is how he conducts himself today.

Well, for three years after his release, he attended his parole meetings, followed the instructions and rules laid out by his parole officer at the behest of the parole board and was not just a law-abiding citizen, not just a productive member of society and earning a living and supporting himself and his wife, but he was also fighting to help others who are sick.

Even though, by all that's right, he more than paid for his crimes in his extreme length of incarceration for a non-violent crime, he was still being subjected to more control and oversight than released murderers do. However, not only didn't he complain about it, he ended up becoming friends with his parole officer.

These are not the actions of a hardened criminal who needs to be supervised. These are the actions of someone who has decided that he never wanted to spend another day in jail and acted accordingly.

So again, what's the problem?

Robert's original parole officer and friend passed away and he has a new parole officer now. This new guy is operating under a different set of rules than the old parole officer. Despite the fact that there were never any problems with his previous one and he had been granted early release on the recommendation of the original one. The new one has imposed heavy new restrictions on Robert, made false claims about him and is preventing him from carrying on not only his livelihood, but also from saving the very lives of him and his wife. Despite the fact that Robert had a doctor's recommendation and his former parole officer's permission to use Cannabis, the new parole officer used the fact he tested positive for Cannabis to punish him and is threatening to send him back to jail for using it.

This is an example of the law being wrong.

I can't tell it nearly as well as Robert himself can, so I offer his own words about it.

Two Faces of American Justice

By Robert Platshorn

After spending almost thirty years in prison for importing marijuana, I met Tony, who was to be my parole officer for the next three plus years. A big strong guy in his late forties, he looked like the kind of fed I had hoped never to see again. For the first year it was touch and go. Tony trying to convince me he was a decent human being and me confident that he was just waiting for a chance to put the Tuna back in the can.

Score one for Tony! For the first year while I was finishing my book, he allowed me to travel to earn a living in my old profession as a pitchman. He visited once a month, gave me the required number of piss tests and tried not to interfere with my reclaiming some sort of life for Lynne and myself in our few remaining years. I am now almost seventy.

At the end of that first year, he announced that the piss tests were over and wished me luck promoting my memoir, Black Tuna Diaries, and the upcoming documentary telling my story, Square Grouper. Tony knew that I would have to travel frequently to book signings, speaking gigs and promotions for the movie. His only admonishment was that I let him know when and where I had to travel, and to call him when I returned and let him know that I was alright.

I came to learn that his concern was genuine. Tony never felt it necessary to issue formal travel permits and just wanted enough paperwork to cover his butt.

Two Deadly Diseases

For the next two years, life on parole was tolerable. After serving two years on parole, Tony put me in for early release. It took months for the US Parole Commission to respond. In the meantime, both Tony and I were having serious health issues.

He developed Aspergoliosis, a deadly lung infection. I was paying the price for years out on the ocean fishing. Skin cancers were popping out on my body like zits on a teenager. At each monthly visit, Lynne and I could see Tony wasting away as a result of massive doses of chemo that didn't seem to be working.

By then, we regarded him as a friend and we were concerned. He in turn was concerned about a large carcinoma that had been removed by a Mohr's procedure from my right calf and refused to heal.

When it was clear that the wound wouldn't close and the cancer was returning on the edge of the wound, I told Tony that I intended to try treating it with cannabis oil. He raised no objection! I already had a doctor's recommendation and had managed to get a small amount of real Phoenix Tears while visiting Michigan, where it is legal.

After three days, the wound was growing fresh skin, not scar tissue, and the cancer began to turn black and die. On his next visit, I showed Tony the healing progress. He was pleased for me.

Lynne and I on the other hand, were becoming worried about Tony. He was on six courses of Chemo a day and looked like a wasted cancer victim on his last legs. He couldn't eat.

That's when our discussions about medical marijuana got really serious. He knew it could solve his eating problem, strengthen his immune system and relieve the awful pain caused by his chemotherapy. Lynne and I urged him to give it a try.

He told us that he wanted to try cannabis, but feared it would show up in his regular mandatory government piss tests and cost him his job, and his pension. He was worried about his wife and two daughters in college. He just couldn't take a chance, but encouraged me to continue using cannabis oil to treat new skin cancers. Those decisions may have cost him his life and me my freedom.

Over the next few months as he got thinner and weaker, I would show him the new skin cancers and the quick results I got from the oil, but we couldn't convince him to do the one thing he knew might stop his decline.

During that time my early parole release came through, but I continued to see Tony until he could get the US Parole Commission to give me a final release from all my parole obligations.

The last time I saw Tony, was just before I left for the Denver High Time Cannabis Cup and the Arizona Medical Marijuana conference run by Patients Out of Time. Tony was leaving for a two week stay in a Jacksonville hospital.

When I returned to South Florida after the two week trip, as usual I phoned Tony to let him know I was back and OK. His cell phone wouldn't take a message. The box was full! Both Lynne and I knew something had to be wrong.

A few days later Scott Kirsche, who I had never heard of, called to inform me that he was my new Parole Officer. A few days later he showed up at our door with a piss cup in hand. Lynne took one look at him and labeled him the fascist PO from hell.

He made it clear to Lynne and I that he was testing me because he knew all about my thirty five year old case and was sure I had been smoking pot all along. Furthermore I was to cancel all travel outside the southern district of Florida. The test of course was positive. And so began our descent back into hell.

Without travel, eighty percent of my income was gone in an instant. Book signings and speaking fees had provided the bulk of our income. Almost all required travel to cities around the country. Lynne who has had a heart attack, and suffers from COPD and severe neuropathy, can no longer afford several of her more expensive meds. Rent and groceries are about all we can now afford.

It gets worse.

I was ordered to travel to the parole office every Friday for a month, for a piss test. Not great news for a seventy year old man with a thirty five year old pot case. Great use of government resources!

At the third visit, I was forced into a meeting with Kirsche and his boss Frank Smith. My attorney, Mike Minardi, was barred from the meeting. Smith announced that Kirsche had been acting on his orders. It seems they both are looking to earn government brownie points at my expense and the expense of my wife.

At that meeting the two of them "ordered" me, among many other things, to stop using cannabis oil to treat my cancers.

Later, after granting me permission to travel to Chicago to address the Annual Meeting of the American Bar Association, something I dearly wanted to do and had earned by three years of hard work and a clean record.

A week later, the permission was withdrawn with this chilling admonition, "You are no longer permitted to travel to promote the legalization of cannabis, without the express permission of the US Parole Commission in Washington D.C.

Oops! Do ya think someone might be stepping on the First Amendment to stop The Silver Tour and Grandma?

Although both Lynne and I had asked repeatedly, no one will tell us what was happening with Tony.

Kirsche and Smith had reported me to the parole board for traveling extensively without permission. A serious violation! And not true!

Despite repeated requests from my attorney, Smith and Kirsche refused to even make a phone call to confirm my travel arrangements with Tony while he still lived. They knew I would not be able to disprove those allegations with Tony gone.

We eventually learned that Tony had passed away about a month after my return.

I used the forced confinement to finish our TV show, "Should Grandma Smoke Pot?". If you've seen the show, you know Grandma has now educated millions of seniors about the benefits of medical marijuana and brought them into the fight for rational cannabis laws.

I had six months cancer free thanks to cannabis oil and Anthony Gagliardi, a good man who thought I was entitled to a decent life after spending thirty years in prison for a plant. In the past three months without cannabis oil, the cancers have returned in force. Face, arms, chest, and torso. Yesterday my dermatologist biopsied half a dozen squamous and basil cell carcinomas.

I do not want to be sliced and diced anymore. Not when I know for certain that there is a better and safer way to have them gone in just days. That I can no longer afford the Medicare surgery co-pay it seems is irrelevant. I am denied the right to choose!

If I catch dirty urine, Kirsche and Smith will be delighted to send me back to prison.

I'm sure it never entered their minds that Lynne can no longer survive on her own, and at my age, I might not live to make it out again. No matter!

Unless the Parole Commission or the courts decide they have had their pound of flesh, I will never again spend one day as a free man who can make his own choices. But that doesn't concern Kirsche and Smith. They claim they are just doing their job.

I seem to remember that lament but never heard it from Anthony Gagliardi who believed it was his job to help people like me to succeed in civilian life after leaving prison.

How does it make sense that the government and law enforcement are spending resources to harass this man? How does it make sense that they can overrule a doctor's recommendation? How does it make sense to stop a man from doing legitimate work helping and educating others? How does it make sense to restrict his ability to support him and his wife and make him dependent on the state?

Most importantly, how can the government stop him from using a plant that is saving his life?

What happened to supporting people in rebuilding their lives as a law-abiding citizen after they have paid for their crime? This seems more like, deny him his medication and hope he dies before he makes much more noise.

I urge you to go visit Robert's web pages and show him support.

The way to make this right, beyond legalizing Cannabis, is for the government to release him from parole and all conditions. I am including addresses in the appendix where you can write the Parole Board and show support for Robert being granted a full release from parole and being allowed to live out his final years in peace.

That isn't the end of Robert's story though. Even while being restricted as he is, he is still doing his best to promote the healing aspects of Cannabis and pushing to get it legalized in Florida.

Robert was finally released from all parole in September of 2014 and in November; High Times Magazine brought him to Amsterdam and awarded him "Freedom Fighter of the Year" for pioneering work educating seniors on the benefits of medical marijuana.

His main site can be found at;
http://thesilvertour.org/

Chapter 11: "Millie" and her son's Crohn's

This identity of the people in this story is hidden as they live in an American State that is still prosecuting Cannabis users despite all the evidence supporting it. The details of the story however, are not made up.

For three years, Millie was told that her son was lying about having pains in his stomach. When routine tests didn't show anything, the doctors didn't just tell Millie this, but they did it in front of her son and told other people the same about him. Regardless of anything else that happened, that is really one of the more disgusting aspects of this for me.

To accuse a child in pain of lying is to add shame and a fear of seeking help. No matter what happens in this boy's life from now on, it will be colored by the doctor's accusations. If he needs help for something, the first thing he will have to do is overcome the fear of being accused of lying by those that are supposed to help him.

Of course Millie knew he wasn't lying, after all, she was the one who saw him turning away from playing with his friends, being curled up in bed crying in pain, not to mention the effects of not being believed about it in the first place. She tried over the counter medications which helped ease and mask the pain, but obviously weren't doing anything to actually deal with whatever the problem was in the first place. It was so bad that she had to officially pull him out of school and home school him seems with the doctors insisting he was lying and there was nothing wrong with him, the school was threatening to take them to truancy court for his missed days. She had to give up her full time job and the benefits that came along with it in order to make sure her son was still getting an education. All because the doctors insisted that he was lying and faking.

Unable to take it anymore, Millie returned to the doctors and demanded an appointment with a Gastroenterologist to see if she could find help for her son. The first test he did discovered three bleeding ulcers. Her son was prescribed some medication for them, which like the over the counter ones, helped mask the problems but not only weren't doing anything to actually fix them, but were also causing negative side effects both physically and psychologically.

Obviously unable to simply accept that her son was sick and in pain or just lying; Millie started researching his various symptoms and issues to see if she could find something that may help him. This led her to demand food allergy tests, which showed severe allergies to both milk and eggs. Again though, although cutting these things out of his diet helped, it still wasn't fixing whatever the underlying problem was.

By age 13 he had been to the emergency room several times over this, so despite the objections of her doctors, she got them to order more tests, including an ultrasound of his abdomen. That test showed that he was suffering from severe gallbladder disease and had giant gallstones. Emergency surgery removed his gallbladder as well as some liver tissue due to the amount of accumulated damage of being ignored and given the wrong medications for so long.

Another year of even more pain plus severe diarrhea and yet another doctor and he was finally diagnosed with Crohn's disease. With a proper diagnosis in hand, one would think that this child would now be on the path to recovery. Unfortunately, like many Crohn's patients, not only did the conventional treatments not work, among other things, some of them were causing her son to hallucinate, once for twelve days straight.

One of the things that Millie came across in her research was mention of people using Cannabis to treat their Crohn's. She had originally dismissed it due to the misinformation that has been spread around for the last eight decades. However, after watching her child hallucinate for almost two weeks due to a prescribed pill, comparing with her own experiences having smoked Cannabis when she was younger, she was willing to at least let her son try it and see if it helped.

When all else fails, what choice does a parent have but to at least try?

So, Millie became a criminal for the sake of her child. After a lot more research online, she bought a vaporizer and went out and tracked down some Cannabis. She then sat down with her son and they discussed what she had learned and why she thought he might want to try this. Keep in mind we aren't talking about a typical teenager who is out partying and having fun with his friends who may have tried Cannabis for recreation, this was a child who was having his life stolen away day after day with little chance to experience anything but pain.

Within moments of his first time using the vaporizer, she saw him start to relax as the pain faded for the first time in years. Since that time, he has gone from taking 8 harsh prescription drugs a day, to a single mild one along with the Cannabis.

Sadly though, the fact that he had to now lie to his doctors and everyone else around him about what is helping him has caused him to experience guilt over it. It's especially worse as he is now doing what they originally accused him of; lying to them. This has caused him to see a therapist to help deal with his emotions, unfortunately though, he can't even tell them the truth for the same reason he can't tell anyone else. To do so would mark both him and his family as criminals; simply for trying to save his life and get him some relief with a plant.

It's now been two years since he started using Cannabis and the Gastroenterologist is amazed by his recovery and progress even though he has no clue why it's happening. This is the very nature of the problems that the suppression of this natural plant has caused. Not only are kids being made to suffer due to doctor's unprofessional behavior and errors in diagnosis, but they are further made to suffer for finally finding something that helps them.

Of course, using Cannabis through a vaporizer isn't as nearly as effective as eating oil is, but making the oil adds a whole new level of stress to the situation for them and they had been afraid of trying it. They recently have made the decision to try some though and I will be happy to update their story in the next edition.

Chapter 12: Organizations

"The essence of community, its heart and soul, is the non-monetary exchange of value; things we do and share because we care for others, and for the good of the place."
- Dee Hock

As well as highlighting the personal stories of people around the world fighting to fix the laws and legalize this plant that heals so many people of so many things, I thought some time should be devoted to some of the organizations that are doing the same thing.

Not all of these groups are necessarily fighting for the complete legalization of the plant. Some are fighting for specific strains or products to be allowed under some circumstances, some want outright legalization and some are in between. One of the things that I hope to accomplish through this book is to get everyone to understand that there are too many different aspects to this plant to support just one of them.

Some of the organizations are just a single person who has been working to change the laws in the manner they think is best, sharing information, others are made up of a couple people who are helping others directly and some of them have been doing this for a very long time. They are all doing their part to make this plant legal again.

In all cases, check them out and see what you can do to help them, because they are trying to help you.

Here is a list of some of the organizations that are involved in this fight to turn over the ancient leaf;

National Organization for the Reform of Marijuana Laws
http://www.norml.org
@NORML

European Coalition for Just and Effective Drug Policies
http://www.encod.org
@encod

Marijuana Policy Project
https://www.mpp.org
@MarijuanaPolicy

Law Enforcement Against Prohibition
http://www.leap.cc
@CopsSayLegalize

Drug Policy Alliance
http://www.drugpolicy.org
@DrugPolicyOrg

Common Sense For Drugs Policy
http://www.csdp.org
@DrugPolicyFacts

Stop The Drug War
http://stopthedrugwar.org
@stopthedrugwar

Drug Sense (MAP)
http://www.drugsense.org
@DrugSense

This is by no means meant to be a comprehensive list, look around your area, there is most likely someone who has organized a group of some type, or who are just waiting for you to do it yourself.

In the next couple of chapters, I'll highlight a couple of more organizations that are involved.

Chapter 13: Rick Simpson and Phoenix Tears

A lot of people are running scams using Rick Simpson's name. There is only one source for information about Rick, his website at www.phoenixtears.ca If someone tells you they are working with him in some way, and you can't find mention of it on Rick's site, they are lying.

I am not going to get into a long background story on Rick, as you can find out anything you want to know about him in other places. I will sum it up though.

He grew some plants, made an extract from them to condense the active ingredients into a compact form (that is referred to as oil) and used it to cure his cancer as well as many other issues he was having, with no harmful side effects. He grew more plants, made more oil and gave it to people around him who were suffering from various things.

Some of those people, much like Rick himself, were told by doctors that there was nothing more the medical system could do for them and they should go home and get ready for death in the coming weeks or months. It healed them as well.

He wasn't doing this on the sly, or in some shady back room. He was very public about it all, including exactly how he grew the plants and turned them into oil to give to sick people. He even shared it all online for free, so anyone could do it for themselves.

Phoenix Tears was born.

On his site, you will find all the information you need to make your own oil the same way he did. There is no charge for it, there is no requirement to sign up for anything, just go and read it yourself, watch the videos of the process, testimonials from people who have done it and how it has helped them. This was actually the place I first found out about the oil and where I learned how to make it for myself to cure my own cancer.

Now, before I go onto more about Phoenix Tears itself, I want to point a few things out that some people seem to keep missing. He did not create a special strain for this, he grew one of the normal strains that existed already. He did not create the method for extracting the oils, it has been and is being used for many different plant extracts. He is however, as far as I know in recent history, the first to be promoting it as a cure for cancer and providing the method to do it. I know for sure that when I first started searching out ways to deal with cancer, his was the single name I kept coming upon over and over. These days, there are many different people, both privately and commercially working on aspects of this, but Rick should be recognized as the originator of the cannabis oil cancer cure method. This is important to point out, because I have seen many people and especially corporations suggesting that they have something that nobody else has and theirs is the only thing that should be talked about.

Here's the thing; ignoring the legal status, almost anyone can grow this plant and make the oil themselves. It will help the majority of people who do follow the instructions Rick provides on his site, for the majority of things that may be affecting them. It should be the first thing anyone tries as a solution. I say this in spite of the fact that you can find studies that suggest there are some issues that it may not help for, or even worse, suggest may worsen a condition. If it doesn't work, then by all means, try some of the other 'proprietary' methods that people are selling and pushing. It's possible that they may have something that will help better than what we can produce ourselves at home for free, but to assume that the one they are making profit off of is better than the one you can do for free on your own, is wrong. Especially if they are not publicly sharing what their uniqueness is, to the point that people can do it themselves; then even if they may be helping some people now for free, their intent can only be to charge others later for it, once they get the support they need. If that wasn't the case, they would do the same as Rick has done and freely share their information and methods. The fact that they don't, to me anyway, shows that they are willing to blackmail people with their health and very lives in order to make a profit, if not immediately, then in the future.

Keep in mind that this doesn't mean Rick couldn't use some money. He has been traveling the world, helping people learn how to make the oil, both in private and in public settings. He has a book for sale to help support him in these efforts. In it, he tells his own story and lays out in detail how to make the oil and his various thoughts on it. You could also just send him a direct donation just to help him out.

How is that any different then what I was just going on about other people charging (blackmailing was the word I used) for their information? Everything to do with making the oil, is fully public and open already, not just on his website, but all across the internet from him sharing it for so long. You don't have to pay him to help you, he already has helped you. If you are in the position to return the favor, help him help others.

That brings me back to Phoenix Tears and his website; www.phoenixtears.ca. It's more than just a general information site. It is a 'how-to' site for the method Rick used to save his and many other people's lives. He lays out his reasons for choosing the materials and method he did, the dosage to use for serious things like cancer as well as maintenance doses to help combat ongoing health damage from our modern environment.

All for free.

Keep that in mind as you come across people who argue that they have some proprietary material or method that they claim will help people for serious illnesses. Note when the people started doing what they are doing, not when they 'say' they started, but find out when they actually did. Remember that there are people who made oil and promoted it as something that was healthy or fun in one manner or another for a very long time, but they weren't clearly stating how they did it, made it or even laid out how to use it, or limited their claims to 'safe' comments about its healing abilities.

There is something else that I find very important that applies to everyone who has used this plant to heal themselves. Whatever it was that they did, it worked, just as they did it. If it didn't then they wouldn't be alive to be claiming it worked. Saying they did it wrong in the face of their continued life, is blatant stupidity. Worse, making claims that they have harmed people by it, such as some people's claims about Rick's methods, is beyond stupidity. Rick saved people's lives whom the medical world had told they had only weeks left to live and there was nothing they could do to help them. Is his method the absolute best and safest? I don't know, but I do know is effective, safe and freely available for anyone and everyone to try for themselves.

People who argue that they have a better method or materials, that Rick was harming people, that only 'professionals' should make the oil or any claims along those lines, are not trying to help people. They are trying to make money by charging people money to help them and they are using the blackmail of death to do so; or they are just trying to stroke their own ego.

I'm including a couple of the paragraphs from the Phoenix Tears website that cover enough of the info for anyone to duplicate what Rick did.

How much to make and take?

One pound of very dry high quality cannabis hemp bud material will usually produce 55 to 60 grams of high grade oil. This amount of oil will usually cure most serious cancers unless the patient has been badly damaged by chemo and radiation. In such cases the patient can often still be saved, but they will have to ingest much more oil to undo the damage the chemo and radiation has left behind. The average patient can ingest a full 60 gram cancer treatment in about 90 days. But if they have been damaged by chemo and radiation often much more oil will need to be taken, over a longer period of time. Sometimes such patients will require 120 to 180 grams to undo the damage from all the chemo and radiation. Once the patient is cured and all the damage has been undone, I recommend that they continue to take a maintenance dose of about 1 gram per month to maintain good health. A small amount of oil about half the size of a piece of short grained dry rice three times a day is a good beginning. After four days double the amount you are taking per dose and try to continue to do soevery four days there after. Until you have reached the point where you can ingest one third of a gram per dose. Taking the oil in this manner in the beginning allows the patient time to build up their tolerance for this substance. Some people soon acquire a very high tolerance and I always tell patients the faster you can take it the sooner you will be cured. I once had an eighty two year old man who was ingesting 2grams a day, who was still going to town everyday and no one could even tell he was taking it. In cases where people are taking strong and dangerous pain medications like morphine. I recommend that they begin treatment taking doses about the size of a grain of short grained dry rice. The idea is to increase their doses as quickly as possible to get off the dangerous pain medications and let the oil take their place to provide pain relief. High quality hemp oil from the proper strains can stop pain that even morphine has no effect on, also this oil can be applied to external injuries for pain relief in minutes.

My process:

I usually work with a pound or more of bud from very potent high quality Indica or Indica dominant Sativa crosses. An ounce of good bud will usually produce 3 to 4 grams of high grade oil and the amount of oil produced will vary from strain to strain. So you are never really sure how much oil you will get, until you have processed the material you are working with. But on average a pound of good bud will usually produce about 60 grams of high grade oil and sometimes you may even get a bit more. Many people will tell you that the oil should be amber and that you can see through it, in many cases the oils that I produced were exactly like that. But the color and texture of the oil you are producing depends a great deal on the strain and solvent that you are using to produce the oil. So don't be concerned if the oil you produce happens to be darker in color, this does not mean that it is any less potent as a medicine.

The process that I am about to describe involves washing the starting material twice with a good solvent such as pure naphtha, to remove the available resin from the plant material. Naphtha has proven to be a very good solvent to produce the oil and in Europe it is often called benzine. The only solvents that I have direct experience with are ether, alcohol and naphtha. Ether is my personal favourite and it is a very effective solvent, but it is expensive and can be quite hard to get. I think the use of ether is better suited for closed distilling devices since it is very volatile and its fumes make it a bit dangerous to work with. Alcohol is not quite as effective as ether or naphtha as a solvent, since it is less selective in nature, but still it does work well. Alcohol will dissolve more chlorophyll from the starting material and due to this, oils produced with alcohol will usually be more noticeably dark in color. For a solvent to be effective it should be 100% pure and 100% pure alcohol is expensive and can be quite hard to find. Naphtha on the other hand is quite cheap to acquire and is usually not too hard to find. Many paint suppliers sell pure naphtha as paint thinners, so for the most part it is quite easy to get and next to the use of ether it is my solvent of choice. All these solvents including alcohol are poisonous in nature, but if you follow these instructions solvent residue in the finished oil is not a concern. When you are done processing the oil after it cools to room temperature, it is a thick grease rather than an oil. The finished oil or in reality (grease) is about as anti poisonous as you can get. Even if there was a trace amount of solvent residue remaining, the oil itself would act upon it to neutralize any harmful poisonous effect. I don't recommend the use of butane as a solvent to produce this medication, since it is very volatile and would require the use of expensive equipment to neutralize the danger. Also using butane to produce the oil does not decarboxylate the finished product, so oils produced in this manner would be much less effective for medicinal use.

The starting material must be as dry as possible; it is then placed in a container of good depth to prevent the oil solvent mix from splashing out during the washing process. Once the starting material is placed in the desired container it is then dampened with the solvent being used, be sure the area you are working in is well ventilated and there are no sparks, open flames or red hot elements in the area. After the material is dampened it is crushed using a length of wood such as a piece of 2×2, after it has been crushed add more solvent until the material is completely immersed, in the solvent. Work the material immersed in the solvent for about three minutes, with the length of wood you used to crush it with. Then slowly pour the solvent oil mix off into another clean container, leaving the starting material in the original container, so it can be washed for the second time. Again add fresh solvent to the starting material until it is once more immersed in the solvent then work it for three more minutes with the length of wood you have been using. Then pour the solvent oil mix into the same container that *is holding the solvent oil mix from the first wash you did. Trying to do a third wash on the plant material would produce very little oil and it would be of little or no benefit as a medicine. The first wash dissolves 70 to 80% of the available resin off the starting material; the second wash then removes whatever resin that is of benefit that remains.*

Use something such as clean water containers with a small opening at the top and insert funnels into the openings, then put large coffee filters in the funnels. Pour the solvent oil mix from the first and second washes into the coffee filters and allow the solvent oil mix to drain through the filters to remove any unwanted plant material. Once the solvent oil mix has been filtered it is now ready to have the solvent boiled off.

Use an inexpensive large rice cooker with an open top that has both high and low heat settings to boil the solvent off the oil. Make sure that the rice cooker is set up in a well ventilated area and place a fan nearby to blow away the fumes as the solvent boils off. Rice cookers are designed to not burn the rice as it cooks and the temperature sensors that are built in, will automatically put the cooker back on the low heat setting if the temperature within the cooker begins to get to high. When producing oil if the temperature gets too high it will vaporize the cannabinoids off the oil and of course you do not want this to occur. That's the reason I strongly recommend the use of a rice cooker to those who have never produced oil before since it eliminates any danger of this happening, if the rice cooker is working properly.

Make sure there are no sparks, open flames or red hot elements in the area while you are filling the rice cooker or boiling the solvent off, because the fumes produced from the solvent are very flammable. I have used this same process thousands of times and have never had a mishap, but for your own safety please follow the instructions, I also caution you to avoid breathing in the fumes that solvents produce. Fill the rice cooker until it is about three quarters full, this allows room for the solvent oil mix to boil the solvent off without spilling over. Put the rice cooker on its high heat setting and begin boiling the solvent off, as the level in the rice cooker drops continue to carefully add the solvent oil mix you have remaining, until you have nothing left. When the level in the rice cooker comes down for the last time and has been reduced to about two inches of solvent oil mix remaining, add a few drops of water to the solvent oil mix that remains. When I am boiling the solvent oil mix produced from one pound of starting material, I usually add 10 to 12 drops of water at this time. This small amount of water allows the remaining solvent to boil off the oil that remains in the cooker more readily. When there is very little remaining in the cooker, I usually put on a pair of gloves and then pick up the cooker and begin swirling its contents.

Until the cooker automatically kicks off its high heat setting and then goes to low heat. As the last of the solvent is being boiled off, you will hear a crackling sound from the oil that is left in the cooker and you will see quite a bit of bubbling taking place in the oil that remains. Also you will notice what looks like a small amount of smoke or steam, coming off the oil in the rice cooker. But don't be concerned this is mostly just steam produced from the few drops of water that you added. After the rice cooker has automatically switched to its low heat setting, I take the inner pot out of the cooker and pour its contents into a stainless steel measuring cup. There will be a small amount of oil remaining in the pot that you will find almost impossible to get out, unless you use something like dry bread to absorb the oil while it is still warm. Then small amounts of this bread can be eaten as a medicine, but remember it can sometimes take an hour or more before you feel its effects. So be careful how much bread like this you consume, because it may put you to sleep for quite a few hours, just the same as the raw oil will do itself. Take the oil that you poured into the stainless steel measuring cup and put it on a gentle heating device such as a coffee warmer, to evaporate off whatever water remains in the oil. Quite often it only takes a short time to evaporate the remaining water off, but also some strains produce more natural turpins than others. These turpins can cause the oil you now have on the coffee warmer to bubble for quite some time and it may take awhile for such oils to cease this activity. When the oil on the coffee warmer has stopped bubbling and there is little or no *activity visible, take the oil off the coffee warmer and allow it to cool a bit. Then using plastic applicators or syringes with no needles, that are available in your local drug store. Use the plunger of the syringes to slowly draw the warm oil up into the syringes and allow it to cool. In a short time the oil will become a thick grease, sometimes the oil can be so thick that it can be hard to force it out of the syringes when cooled. If such a thing happens simply run hot water over the syringe and your doses can then be forced out much more easily. Sometimes a patient will force out too much oil, but if this happens just pull back on the plunger of the syringe and the excess oil can usually be drawn back into the syringe without too much difficulty.*

On average if I have a dry pound of material to work with, it will require about two imperial gallons of solvent, or 9 liters which equals about 320 fluid ounces to do the two washes that are required. If you plan to produce the oil from more or less starting material, simply do the math to determine roughly how much solvent you will require. From start to finish it usually takes me about four hours to accomplish the whole process, then the medicine is sitting there ready to be used. It should also be mentioned that this oil has an extremely long shelf life, if kept in a cool dark place for storage.

I think these instructions should make producing this oil quite easy for anyone, but before you start make sure that you have everything you will need to do it properly. At first it may seem daunting for some to try to produce their own medicine, but in reality this process is extremely simple. All you have to do is carefully follow the instructions and after you produce this medication a couple of times, you will find that it is not much harder to make than a cup of coffee. Once you have produced your own medication it takes all the mystery out of medicine and you no longer have to rely on doctors in most cases, for now you are your own doctor. Welcome to the world of real medicine, medicine that does no harm and is effective for practically all diseases and conditions and a wonderful natural medication that you now know how to produce yourself.

Best Wishes and Good Health,

You can also find videos online if you search for "Run From The Cure", or "Rick Simpson Oil".
www.phoenixtears.ca

Chapter 14: On Doctors and the Medical System

"The doctor has been taught to be interested not in health but in disease. What the public is taught is that health is the cure for disease." - Ashley Montagu

I am not anti-doctor or medicine, I am just against a profit driven health care system.

What's the difference you ask? Pretty much everything from the way patients are dealt with to the treatments chosen or ignored. When profit is involved, especially with corporations, there is no other concern besides profit. Now you will of course have many people who profit off corporations and business enterprises that will disagree with me, but the simple fact is that if they were trying to help people, they would not be getting rich while some people suffered from something they could alleviate if money and profit wasn't involved.

Most of us can agree that it is wrong to walk up to a person who has never harmed you or anyone else that you know of and shoot them in the head. We wouldn't think it was okay if they gave us a chance first by saying, give me $1000.00 and I'll let you live; in fact we would consider that another type of crime. Yet when a doctor says, "You are going to die unless you pay me $1000.00 for this medication", we somehow think that is okay.

The person in both cases has the exact same choice, with the exact same result, pay $1000.00 or die. Some corporations and people like to try to make it sound better by saying that if someone is truly in need they will give it to them cheaper, but that is just a public relations stunt. If you have 10,000 people that need the same drug who can't afford it and you give it to one of them while still demanding the other 9,999 find the money or die, it means nothing except to that one person. Sure, they will be grateful you helped them and that they get to live, but it doesn't make you a good person.

So, again, I am not against doctors or health care, I am against it being used to profit at the expense of people's lives.

Of course, if the profit driven corporations blackmailing people with their very lives was the worst of it, we would still be in a pretty good position, comparatively speaking. Unfortunately, that quest for profit at all costs, means they do worse than just not help people if they can't pay. They lie about the products and services they sell in order to increase their bottom line.

This isn't a new idea or a crackpot conspiracy theory, it is based on the very open fact that corporations in every industry including the health and pharmaceutical ones, have been caught doing these things. Worse than that, they have spent millions of dollars lobbying to get protection for themselves when they do them, plus more millions fighting or hushing up victims of their policy choices.

These things are not a secret; we see them in the news every day.

So, having said all that, I will now point out why I felt it was necessary to write this chapter about why none of that means I am against doctors, health care or medication.

There are people, both doctors and in the business world, who honestly do care about others and want to help. In some cases they do charge money for what they do in order to cover their costs. What they don't do, is charge so much money that they can afford to work one or two days a week and still be in the upper income brackets of our society while ignoring those who can't pay.

You will find these people in all sorts of places, doing the best they can in a system that is designed to funnel profit to the top, regardless to the cost to those below. They go way above and beyond what most people do in order to help, sometimes at the risk of their very own lives. These people should be appreciated.

However, even among them, a lot have been trapped into a profit geared system that allows little leeway for being an actual nice human being. They will always try to listen and understand what is wrong before simply ordering another procedure or pharmaceutical product and if their patient can present information that counters something they think they know, they at least have the sense to look into it in order to make an informed decision rather than repeating a corporate tag line passed down by salespeople.

Now, when it comes to medicine, the way I look at it is pretty simple. If it is something that's sole purpose is to help, it is medicine; if it's purpose is to make a profit, it is not medicine, it is a commercial product.

This beyond anything is why I will almost always prefer a natural product that someone can grow and make at home, over something that is being sold for profit.

However, that does not mean that I automatically think all commercial pharmaceutical products are unnecessary or solely for the purpose of profit. Like all things, they should be researched and the person who is in need should make an informed decision. Of course, that is hard when the very corporations that make them lie about both their negative side effects and actual main effects, but that just means a person has to be careful and thorough when researching.

As for doctors; there are many things that doctors do that are absolutely necessary and life saving. Things that without them, people would needlessly die.

Just recently there was a car accident where a baby's neck was broken. An internal decapitation is what it was referred to as. If it wasn't for doctors, nurses and the profit based system that spawned them, that child would be dead today. Instead, he should make a full recovery. That's a good thing and it isn't an isolated incident. Doctors and the system do save many lives every day and that should be recognized.

They just shouldn't be considered gods, or infallible sources of information. Like everything else, the pros and cons of it all must be examined and balanced by the person whose life depends on it.

Why do I feel the need to point this out? Because someone who read a lot of my writings said to me the other day that doctors should be considered the enemy and anything they say should be ignored.

To me, that is just as bad as someone saying everything they say, should be accepted.

Chapter 15: Conclusion

All seed-bearing plants I give to you - God

Now that you have had a chance to read the whole book, get an idea of the concepts at play and see the personal stories of people who have used the Cannabis plant to heal themselves and those who are helping to heal others, I hope to be able to summarize a lot of this into something that we should all be agreeing on: the complete and total legalization of this ancient plant. If the governments want to make regulations over the commercialization of it and gain tax revenue from it, I say let them; as long as they allow those who wish to grow, use and share it on a personal basis to do so, just as we can grow, use and share our food, without restriction or harassment.

There are too many studies by non-biased sources these days, not to mention the thousands and thousands of years of history of our use of this plant that proves there are not only no harmful effects, but our bodies have evolved to use its properties in the maintenance and recovery of our health.

As for its recreational uses, as it has no harmful effects, doesn't cause the same social issues as alcohol and when not restricted, causes no financial or legal hardships on its users, it should be hailed as the best choice for everyone.

What I hope to see is that all those who are currently struggling to change the laws where they live to allow certain aspects of the plant to be legalized, pull together and fight for the complete legalization of it worldwide. It is the best and most effective gift we can give to our future generations.

When you stop and think about all the benefits that people are reaping from this plant, that we could all be growing in our backyards or on our windowsills, you have to wonder what the future holds for us all. Imagine one where instead of going to the doctors and spending thousands of dollars on things to try to cope with something that already went wrong, you could be using something as part of your daily diet and health routine that eliminated the majority of things that ail us all?

Imagine a medical system that wasn't being overwhelmed by patients and was able to deal with the things that nature hasn't already provided the answer for.

We would have societies where everyone was blossoming, instead of spending the majority of our resources to make a few people richer. The opportunities to explore the avenues of knowledge that are being ignored now, in the rush to profit off of people's very lives, would allow us to start solving the other problems we face, such as healing the planet. Of course, the Cannabis plant also has a huge role in that aspect as well.

Let's turn over the ancient leaf.

###

Thomas Henry Crinstam

Hello, I'm Tom, the author of this book. I really appreciate you taking the time to read it and hope that at the very least, it gave you some new perspectives on things around you.

I wish I could say that I would be happy to hear from you and you should contact me if you wish to, but unfortunately, that isn't always beneficial to anyone involved. Regardless of the labels, treatments or sometimes even reality, I can be a very difficult person to get along with, which is sort of sad as I really do love people and love connecting with them.

If you think you can handle it, look me up and say hi.

I wish you the best in life.

Tom

Facebook: https://www.facebook.com/thomashenry.crinstam
Twitter: https://twitter.com/THCrinstam
Blog: http://www.thcrinstam.com/

Final Note

An end is just a new beginning.

The first time I released this book, we were in a pretty dire situation here. Besides the legal aspects which kept me from being able to produce my own medicine which I desperately need to save my life, we had suffered a series of blows that left us planning on repairing an old motor home and making it suitable for year round living. That was literally our last resort before we ended up homeless.

However, a weird series of events happened that changed everything that along with the election literally reshaped my entire world and future.

It started with the owner of the house we were renting losing his job out in the oil patch and giving us notice that he needed his house back here. Instead of having until spring to raise the money and do the work to get the motor home ready, we had less than two months to find a new place to live, which would have been around the start of November.

So we started desperately looking for another solution; after all, living in a motor home in Canada during the winter isn't something that should be done unless you are ready for it. It would have been better than being homeless completely, but only by a thin margin.

We heard about a house that might be available for rent or sale not too far from where we were, which meant among other things, that Nicole could keep her job and we would still be close to her family, which was the original reason we had come here. So, we took a drive to see if we could find it.

Find it, we did. There was a For Sale By Owner sign in the window as we pulled into the driveway with contact information on it. We looked around at the place and were really amazed by it. The sign said 7 acres, 2 bedroom house. Nobody was there, so we couldn't get inside at that time, but we walked the property, checked out the outbuildings and pen areas, the garage is big enough to park in and still leave room to work on other things, there's a small pond in the back and no real immediate neighbors.

First, Nicole called and asked for more information and tried to feel them out to see what type of deal they were looking for on it. After all, we lost everything over the last couple of years, we didn't even have enough to pay first and last month's rent on an apartment, let alone buy a house. Next, I emailed the owners and was blatantly honest with them; I laid out our situation and what we could do in terms of a deal. Honestly, I didn't have much hope for it as even on a really good deal, this place was more than we could conceivably consider.

However, a deal we did make; and it was a pretty sweet one. The main part of it being that we didn't make a down payment just started making monthly payments. To alleviate the worry of those who like to worry about such things; yes, it was a deal done through the lawyers and is an outright purchase.

Unfortunately, even on the great deal we made, we will barely be able to make ends meet just from a living expenses aspect. That isn't different than we were before of course and at least now we are paying down the balance a little each month on something we own. The problem is that I still need medication and we still need to be able to get things set up to reduce our future costs. Everything from getting a greenhouse and gardens set up, to making sure we have the tools and equipment to do what needs to be done is beyond what we have. So, our fundraiser and other efforts go on.

Why should anyone help us to pay for a home? I've asked myself that from the start and I think the best answer I can give is that I don't know.

I could point out that I have always given what I could of myself to help others, even to the point where it has harmed me; but that would negate the reason I did it, which was because they needed help and I wanted to help them. It wasn't a bargain for future favors.

I could also relate my life, from the early start of abuse through so many screwed up things that I somehow survived and made it past; but I look around at the world and I see so many others in other parts of the world that would consider my struggles to be heaven compared to theirs and that just doesn't seem right either.

I could always appeal to your sense of duty to your fellow person, but again, why me over someone else who is struggling along?

Do I think I deserve to be helped? A question such as that hits dead center of a lot of issues that have held me back over the years, mostly related to my early life and the things I survived then.

After thinking about all these things and many more, I realized that it isn't about why I think you should help me. Anymore than the people that I have chosen to help over my life were chosen because of something they did to 'deserve' it. I helped them because they needed help and I could help, period.

I need help; if you can help me, I would appreciate it.

I do want to point something out and make a promise to you first though. Not only is everything I say in this book inspired by a passion to help others in whatever way I can, but everything I do is always done on the same basis when I have a choice.

I make the distinction of 'when I have a choice', because I do have emotional issues that sometimes remove my choices beyond surviving from one moment to the next. At those times, I can sometimes be extremely negative and hurtful with the same things I use to help most of the time, my words. I sincerely apologize to anyone I hurt during those times, I truly wouldn't do it if I had a choice in the matter.

The promise is that any amount of money I ever get, whether that is from selling things such as this book, my calendars or other creations; donations, actual paid work or even if I won a lottery, I will always be trying to find a way to help others in whatever way I can.

To start with, once we get our greenhouse and gardens going, we will try to grow much more than we need. Half of any extra we grow will be sold to cover costs and half will be given to those who need it.

If anyone ever needs a place to stay for a while to escape a bad situation where they feel trapped, we always have an open door. That includes if someone just needs to get away from the stress of fighting a medical battle and wants some time in the country. It's of course dependant on us having enough room at the time, but even on that, as we raise more money, we will build a couple of small cabins to let people use in such cases.

Of course, if the laws are changed as the new government has promised and there are no restrictions on sharing Cannabis, I will also grow as much as I can, make it into oil and give it to those who need it.

And as always, I will continue to speak out on issues where I think my words may be able to help. Even if it may only help a single person.

In closing, I will quote a person who will hopefully come to be regarded as a great man and one of the best Prime Ministers our country has ever had.

"Better is always possible."
- Prime Minister Justin Trudeau

www.ingramcontent.com/pod-product-compliance
Lightning Source LLC
Chambersburg PA
CBHW022125170526
45157CB00004B/1760